EXPLORATIONS

in Photography

Adventures & Advice for
Advanced Amateur Photographers

ERIC K. HATCH

Published by Hatch Photo Artistry, LLC
6812 Stagge Road
Loveland, OH 45140
http://ekhphoto.smugmug.com
513 494 2625
email: info@hatchphotoartistry.com

ISBN: 9780988210707

CONTENTS

INTRODUCTION

This book is called ***Explorations in Photography*** *not because it deals with unknown territory or sloshes through the swamp of camera comparisons, but because it follows my own innate desire to understand what photography is all about, both from an artistic point of view and from a practical one.*

I've never been one of those people who says "yes, Sensei" without asking "and what will be gained by doing this, Sensei?" This questioning came not from disrespect, but rather because I have always wanted to see how things connect, how they work, why they work as they do, and why they matter. My journeys to find answers to those questions as they relate to photography are the explorations in the title.

Writing down what I've learned has always helped me to understand things. Over the years, these writings have accumulated, and now it's time to share them with you.

The book is divided into two parts. The first focuses on motivation, artistic issues, and some explanations of key concepts in digital photography. The second part addresses more practical considerations.

Although I operated a portrait studio for a number of years and enjoyed the challenges of lighting and posing, most of my photography is now done on and off the road. Accordingly, this book emphasizes outdoor photography and travel.

I've had to make some assumptions about you, the reader. Who will benefit from this book? While every photographer can find things of value in these pages, you'll get the most out of the book if you have decided that you are in charge of your camera; it is not in charge of you. You do not just point and shoot. You have also decided that you want to invest the time, energy, and creativity into being as good an artist as you can be.

This book is intended for photographers who want to understand more of the art and science that combine to make photography an addictive, never-ending exploration of yourself, your skill, and the world around you.

PART ONE

Why's and What's of Photography

CHAPTER 1: *Aesthetics*

CHAPTER 1: *Aesthetics*

THE REAL THING — OR IS IT?

Be warned --- I'm going to be philosophical in this article. You'll have to work to follow it closely, but I hope it will be worth it.

There are still people who whine that digital manipulation is somehow cheating, that it distorts the reality of a film camera shot, and it's somehow too easy. I say nuts, for two simple reasons. 1. No camera captures reality. 2. The final photograph depends on the vision and skill of the photographer both composing and taking the picture, and in the "lab" afterwards.

To begin with, let's take a look at the "reality" of a film exposure. A film camera sees only what is there, and sees it only in terms of light and shadow (and color, though that's a separate argument). It does not see trees and hills and soccer balls and smiling Uncle Jack. At best it produces a representation of what it sees; it's not the thing itself. Heck, a picture's not even three-dimensional, let alone four-dimensional (length, breadth, depth and duration.)

The camera – any camera – does not see context. Only humans do that. That's because an exposure is typically very short. Portrait photographers like to claim they capture a moment in time – but in fact, it's usually something around 1/60 of a second or less. Pretty brief moment, and says nothing of what happened right before it or what happens right after it.

It's the before-and-after part that creates a story and gives meaning to human life. Movies tell stories. Photographs only imply them – and frequently they're lying. Take the picture of the little girl and her dolly. She looks so sad! All I did was ask her to "love her dolly" and – CLICK! That's frame 1618.

In fact a few seconds before this shot she was grinning and skylarking about, and a few seconds after it she was smiling again. See frame 1616.

The reality is – at the instant the shutter clicked, the child looked like that one sad frame. But in terms of her overall personality, it's not the reality. It's not related to anything beyond that instant. Anything we might say about this kid is all made-up, an invention. It's a story. Humans make up stories all the time. We do it to bring order and meaning into what would otherwise be chaos and senseless. But it's all fiction.

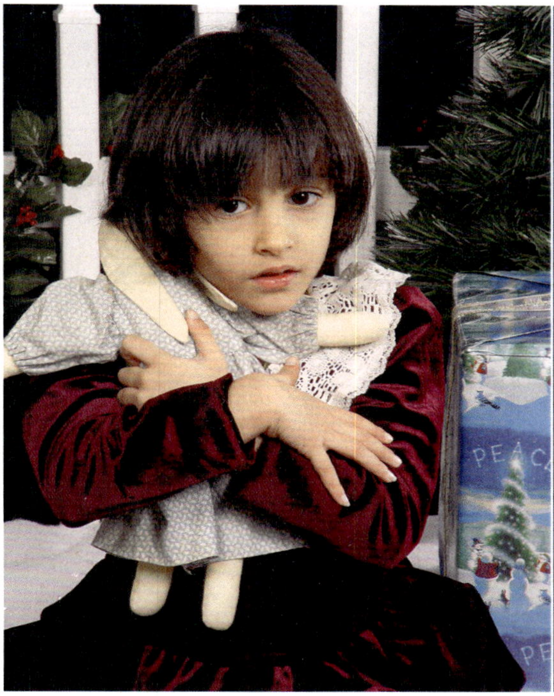

Figure 1-1. *Sad little girl*

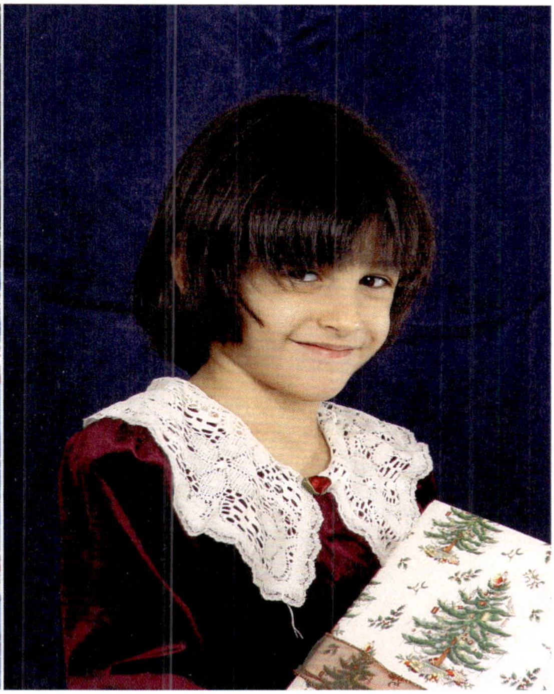

Figure 1-2. *Not so sad, just two frames earlier*

A successful portrait generates a story in us that we like to tell. If the portrait actually generates a story that reminds us of OUR fiction about the person in the picture, it's a really successful portrait. Clients will love it.

So, I ask again. In photographic terms, what's reality? Reality, one more time, is defined by people's experience of life (which includes their experience of science experiments, so nyahh, nyahh, nyahh to you measurers and quantifiers out there).

So, since "reality" in photographic terms is created by the viewer's interpretation of light and dark (captured by the color),what on earth can be wrong with constructing a more artistic or human-oriented version of what the camera saw? No great moral sin is being committed. Removing the inadvertent tree trunk from Sister Helen's head is no lie. Taking it out of there (or blurring it in Photoshop) lets the viewer concentrate on Sister Helen, who is far more important to the photographer and the viewer than the tree trunk could possibly be. All you're doing by taking out the tree trunk is creating the reality you THOUGHT you saw or WANTED to see when you took the picture. The picture now corresponds much more closely to YOUR view of reality, and so is actually MORE realistic than the actual light and dark captured by the camera.

I had to hunt for an example of this particular sin and finally found this one from my first year as a pro. It's embarrassing.

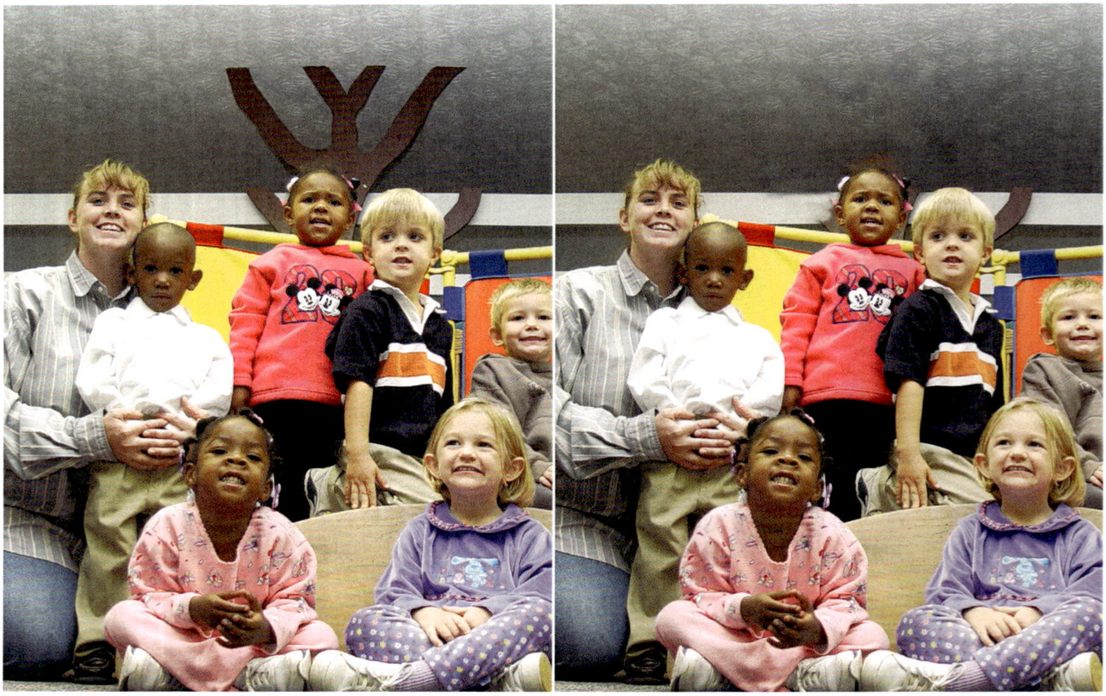

Figures 1-3 (with tree) and 1-4 (no tree). *It's far better to spot the problem before taking the picture. But who's to say which is more "real"?*

The tree trunk appears to be coming from the kid's head for two reasons. The first is that the camera's lens treated head and tree as both in focus with no space between them. The second reason is that the photographer didn't see the tree. He saw the kids, and mentally blocked out the tree. Surprise! It was there all along! Skilled photographers learn to see the tree trunk, and to check the corners of the frame to make sure nothing's there that shouldn't be. So, to finish up, cameras see only a very limited version of reality. All the richness and wealth of reality comes from the photographer – and he or she made it all up in the first place!

Don't be shy. Just admit what you're doing with a film camera or a digital one is only step one in expressing the reality that's in your head .... Then go for broke back in the lab or on the computer!

COLOR VS. BLACK-AND-WHITE

I love color. I'm a fanatic about it. But sometimes color is the wrong way to go. Color can actually be a distraction, putting the viewer's attention in places the photographer doesn't want it to be.

The color version of Figure 1-5 is an excellent photo in its own right. The sky in particular works well, as does the watery sunlight. The black-and-white (BW) version, Figure 1-6, puts our attentions squarely where it belongs, right on the border between shrubs and sky, on the top of the wall. The drama of the scene is intensified in the BW version. It prints beautifully on fine art paper.

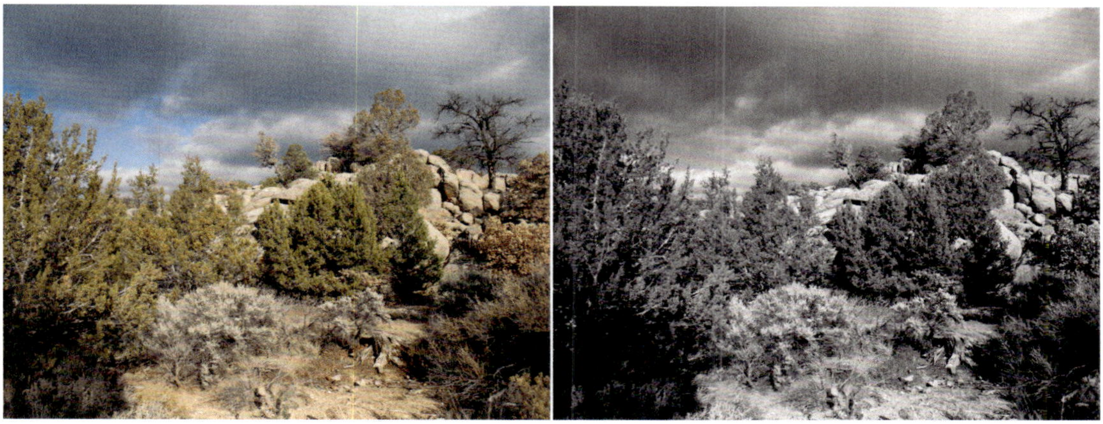

Figures 1-5 and 1-6*, near Dixon, NM. 1-5 is color, 1-6 is black and white. Which has greater drama?*

Even though your camera probably has a Black and White mode, I implore you not to use it. The camera makes decisions about interpreting black and white images and stores them as JPEGs, having literally wiped out all your color data. You can't go backwards! So shoot in RAW (not JPEG – same problem throwing out data) if your camera will let you. Then use image-processing programs such as Aperture or Light Room or Bridge or CaptureOne and convert the image to BW. I use both Adobe's Bridge and CaptureOne; both of them allow you to adjust the luminance (lightness and darkness) of specific colors. This has a huge effect on your conversions, giving you a great deal of control over WHERE in the image areas look lighter or darker based on what color that area was in the original. But you have to watch it – extreme changes introduce a lot of noise (visual static) into the image, and you probably don't want that.

So when do you use color, and when do you prefer BW? You use BW to reduce distractions, bringing out tonality (shading) and shape, which may be lost in a very colorful image.

You use BW when the image is fundamentally monotone (unless you have a good reason for not doing so, like a single colorful element, or "selective color" when just one element is colored and the rest is BW). Again, playing with the lightness and darkness (curve) of each color lets you get the most bang for the buck.

You use color when color itself is the main subject, as in this picture of scarves at a summer festival: Figure 1-7 really does work better in color – in fact, it's all about the colors. But I'm putting it here to show you how flexible black and white conversion can be.

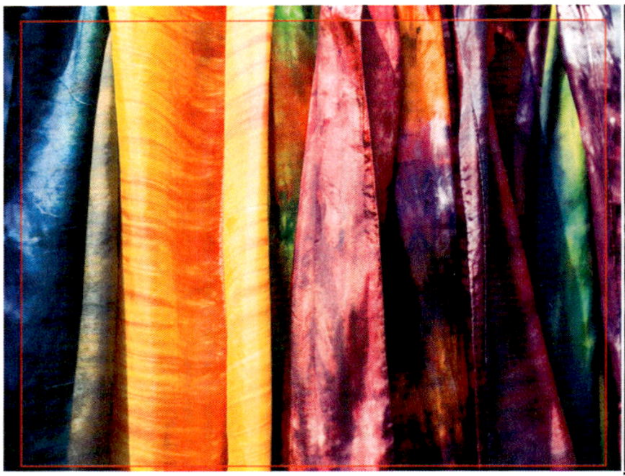

Figure 1-7. *Scarves at a country festival, Hocking County Ohio*

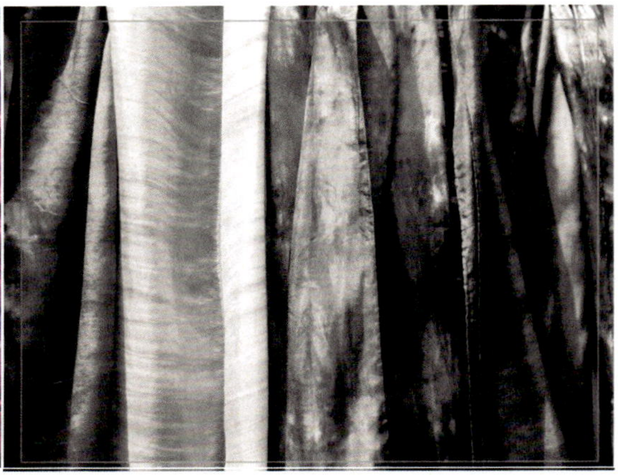

Figure 1-8. *Here's a BW conversion making no changes to the luminance of specific colors.*

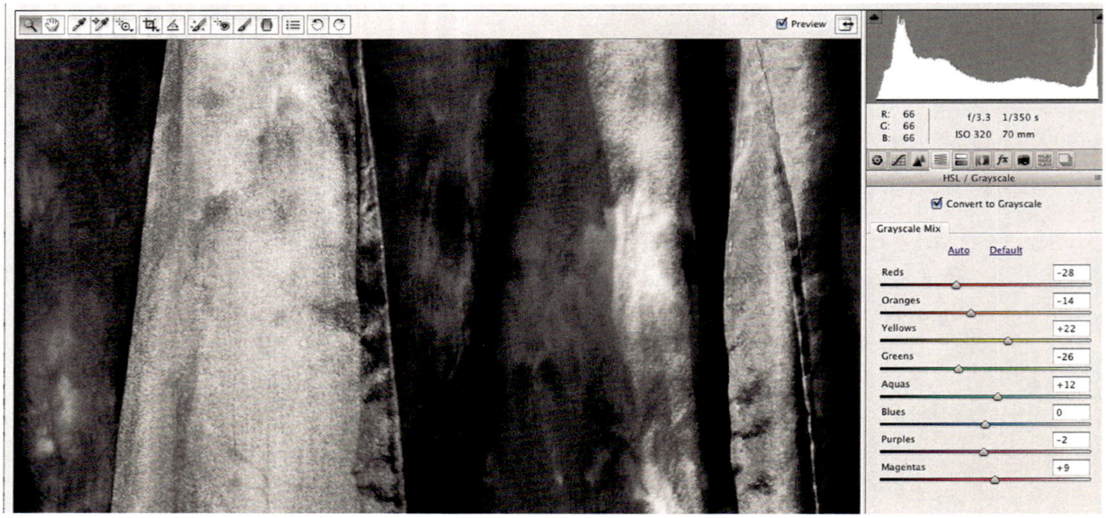

Figure 1-9. *Using sliders in Adobe Bridge (or Lightroom) you can lighten or darken individual colors to alter how the BW version looks. Adjusting tonality of individual colors makes a huge difference in outcome.*

Converting to BW back in the computer by adjusting the tonal values of various colors is a lot like using filters to adjust black and white shots back in the days of film. The same general principal applies, but with a much more subtle range of options.

You can use either color or black and white when you're trying to set a difference in mood. Here are color and bw versions of snow in the winter, taken just at sunset. They're both good pictures; the BW version looks incredible when printed on Lexjet Sunset Cotton Etch paper.

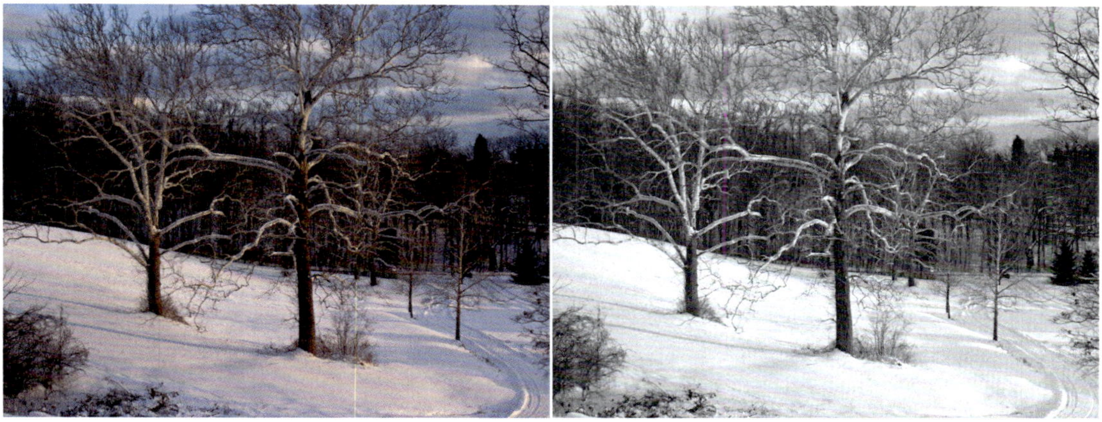

Figures 1-10 (color) and 1-11 (BW). *Sunset on winter's day, Loveland, OH*

All photos print differently on different papers, both in color and black and white. If you're serious about printing black and white, there are entire organizations devoted to mastering the art of black-and-white printing. All I can suggest is play with it, and keep playing until you're happy.

One last thought. When is black-and-white not black-and-white? When it's TONED – as in duotone, sepia, etc.

Toning adds a (small) amount of a second color to the image, which is then printed as a color image. Of course, if you use a warm-toned paper (creamy, rather than bluish), you will also alter the tone of the print, but leave that out of the discussion for right now.

Technically, when you print an image in black-and-white you are reducing the amount of data in the original, and you are limiting the range of the final print. If you add a second color, the total number of shades your printer will print jumps exponentially, so you get much more subtle shifts from light to dark. Not every picture benefits from toning, but many do. Here's an example of a toned and untoned treatment of the same image. The toned version included 3% (just 3%) of a light green color. That's subtle enough so I hope you can see the difference when you read this book. The toned version does not look green; it just looks a bit richer and less stark than the BW version, and you may not notice it at all. But the toned version has a smoother "glide" from light to dark, and consequently has a different feel. There are mathematical reasons for that, as well as practical ones, but they're less important than the magic a little toning can make.

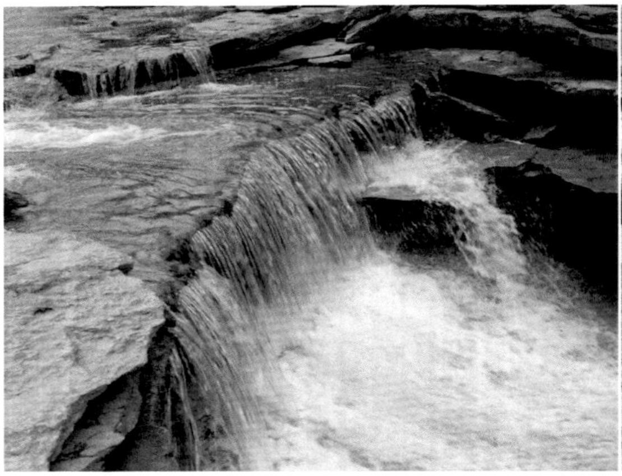

Figure 1-12. *Backbone Falls, untoned*

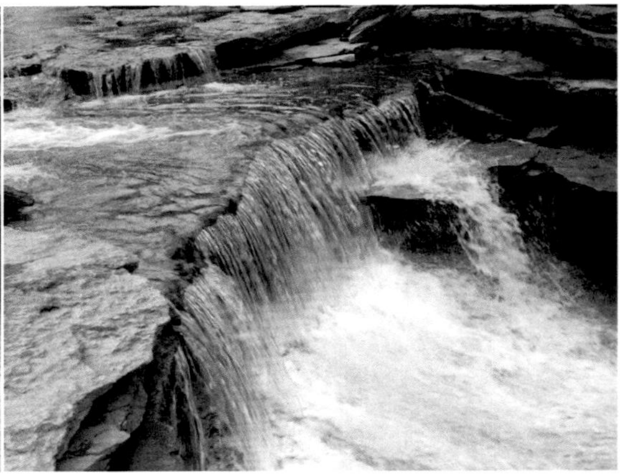

Figure 1-13. *Slightly toned version of Figure 1-12. Look for smooth transitions from light to dark.*

To conclude, there's plenty of room for both color and black-and-white in modern photography. The virtues of BW are mood control ranging from hard and angular to soft and gentle, as well as removing the distractions color often provides. Color, on the other hand, can create a more vibrant look and can dazzle in its own right.

ASPECT RATIOS AND HOW THEY WORK

I've been experimenting with aspect ratios, as a way to understand panoramas and how they work.

Aspect ratio is the ratio of the width to the height of a rectangle. Aspect ratio is notated as X:Y, where X is the horizontal and Y the vertical side. For example, a 6x4 print is in the ratio of 3:2, or (reducing the fraction) to 1.5:1. As it happens, this aspect ratio is approximately the same as that of a full-size 35 mm camera sensor (36x24). It's also very close to the "golden ratio" that occurs frequently in nature and in mathematics: 1.68/1. This ratio dictated the proportions of the Parthenon and a variety of more modern buildings.

This is also the naturally occurring ratio that describes the length of your hand to the length of your forearm, and can also be found in plants and animals: the seed distribution in a sunflower, or the shell of the chambered nautilus. In photography, it's useful in composition (see Chapter 7).
I got curious about the aspect ratio of normal human vision. How do we naturally see things? How does

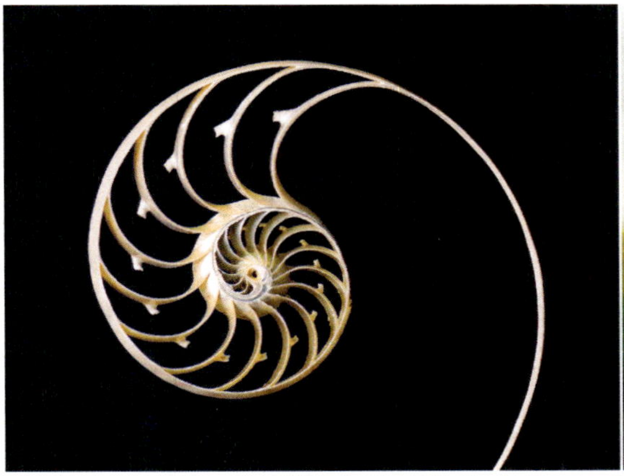

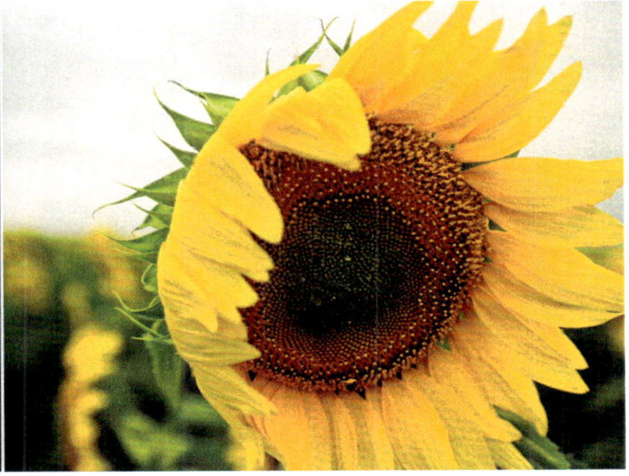

Figure 1-14, *Nautilus shell section, and* ***Figure 1-15,*** *sunflower head, both show occurrence of the "golden ratio" in nature.*

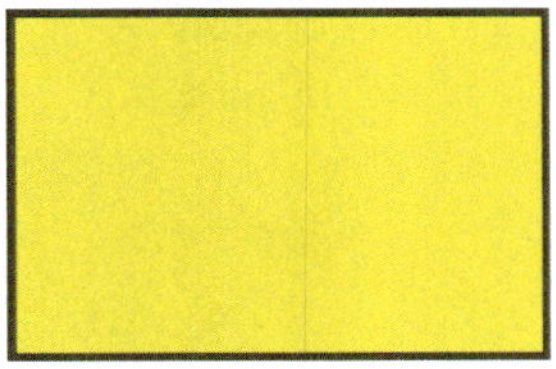

Figure 1-16, *yellow rectangle, is built on golden ratio.*

that affect the way we react to and interpret photos? The closest I could get to working this out was that we normally see in a ratio between 1.5 to 1 and 1.6 to one. This is pretty darn close to 3:2, aka 35 millimeter film format, shown in the yellow rectangle.

What's magic about this aspect ratio? A rectangle shaped like this holds all the information we can see at a glance. You don't have to scan it to get what's going on. If you do keep looking at it, your eye will sweep it from left to right, but basically, you can get the gist of a 3:2 image all at once. This gives images greater impact, makes them easier to decode. Different aspect ratios elicit different ways of "reading" the image.

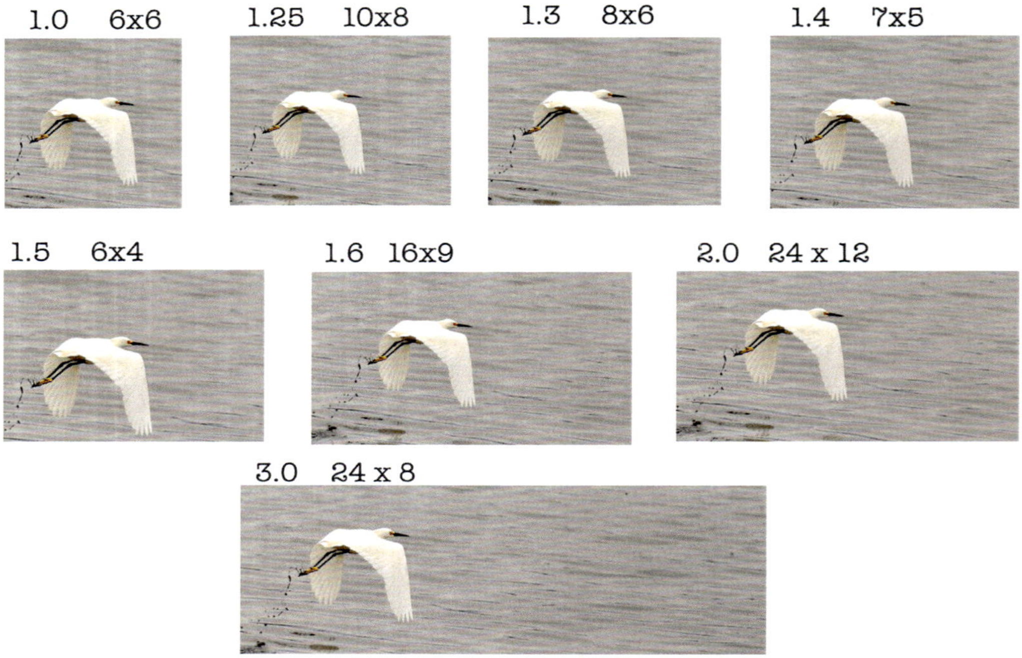

Figure 1-17. *Common aspect ratios illustrated*

As ratios start getting wider, the way we see the picture changes. There's much more decoding going on. This example sheet makes the point. Even though the image starts as a portrait of a cattle egret, as the aspect ratio widens, the picture changes from a portrait to a story – first we see the bird. Then we start seeing bird and water. Finally, the picture is less about the bird itself than the journey it's on – more about the bird flying over water to someplace unspecified.

When we see a wide aspect image, with a ratio greater than 3:2, we have to stop and ask questions --- what's happening here? How does this bit on the right tie into what I see on the left side? In extremely wide examples, 3 or 4 to 1, you simply HAVE to stop and figure out the image. Such images become "puzzle pictures" as often as not. Decoding them (and appreciating the contents) is a slow process, far from the instantaneous grasp of the 8x12, or even 8x10.

Figure 1-18 is a night view of Cincinnati, shown as a 3:2. Figure 1-19 shows a 2:1 version of the same scene. The 2:1 is much harder to grasp – and that's true of simple pictures as well as complicated ones. It's even true at very small sizes, where we can physically see the whole print at once, but we still stop to decode.

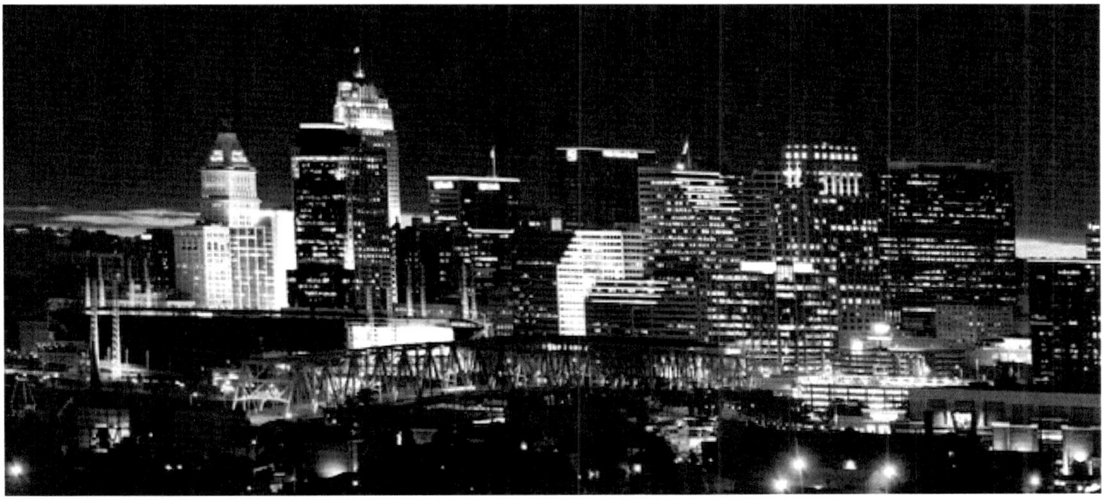

Figure 1-18. *3:2 ratio.*

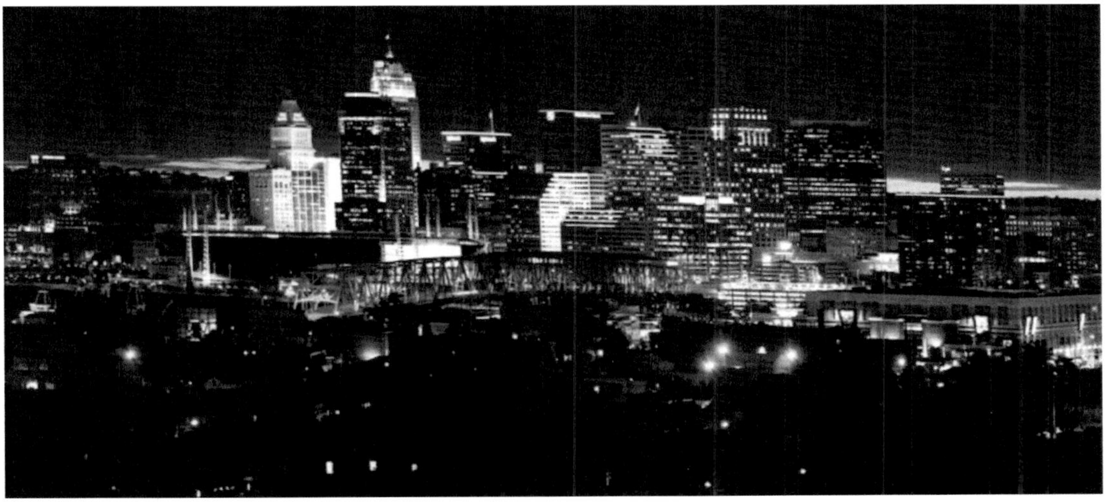

Figure 1-19. *2:1 ratio.*

I'm not making an argument for or against panoramas in this essay. It's just that they work differently, create different kinds of impact and story-telling. In panos you exploit the relationship between material at either ends of the scene. The aspect ratio of the image reinforces the tension and story in the picture.

So have a ball playing with panos – but don't expect them to have the same effect on people as a 4:5 or 5:7 would do at a similar physical size.

WHY BOTHER WITH PRINTS?

I have spent years, literally years, trying to get what's in my eye onto paper. It's a complicated job.

Eyes see one thing (much of it imaginary). Cameras see differently, ESPECIALLY color. A monitor interprets and displays camera data using its own idiosyncratic insights. A printer prints what it thinks the computer wants it to. Result: things slip along the way.

The key is calibration to objective standards, including profiling the paper you use. This takes effort, but it's getting easier to do. My HP Z3100 printer does calibrating and profiling semi-automatically. Expensive? Yes. Worth it? Absolutely.

As I said, it's taken years (and a ton of $$) to get it all working. All to be able to get the best, most accurate image I can manage, onto a piece of paper. And it still takes work: recalibrating the monitor, making new ICC profiles for new papers, re-doing them every 6 months or so when the printer says "out of date! Spend more money redoing them."

So why use prints at all?

Prints – good prints, on archival paper, using archival inks – last a long time, perhaps 200 years, and they don't go out of date every 5-10 years, like storage systems. Remember tape drives, floppy disks, zip drives? Super-8 movies, even VHS video tape? Gone, baby, gone with the wind. Sometimes low-tech has its advantages! (Except that the papers themselves are not low-tech, relying as they do on very sophisticated coatings for their brightness and longevity.)

Unless you update your storage systems and re-do your entire file library, there will come a time when either the storage medium becomes unreadable, or, worse, the data itself can no longer be read. This can be as short as 5 years (any body remember Zip drives?). After a while, newer versions of applications can no longer read documents produced in earlier versions. Five to 15 years, and you're history, dude.

Prints last a while. They don't go out of date, and if cared for, will outlast your great, great, great grandchildren. And you never need new technology to read their data; you're born with the right technology for the job!

Even more important, prints are really high-res. Your monitor simply cannot resolve to as fine a grain as even crummy paper. Even the best monitor is coarse and crude compared to a good paper print. You can tell the difference instantly.

You get more depth, more ability to appreciate fine photography when it's on paper.

And it makes a terrific difference WHICH paper you use. Glossy, metallic, pearl, satin, semi-gloss, matte, baryta (incredibly fine-grained, that), any of the cotton etch or watercolor papers --- the same image will look very different on each of them and look best on one of them.

When I'm restoring photos for folks, I try to match the paper to the sort of paper they brought in. Many really old photos are on lovely, fine-grained, very heavy papers, with a matte finish. Can do. Others are very glossy. Yep, can do that, too. The point is to use the right paper for the image.

Paper does need care. It should be stored in archival sleeves, out of the light. If framed, the print should be matted with acid-free matting and properly mounted. Use UV-resistant glass; museum glass is much better, but it costs the earth. Paper prints do not transmit light, they only reflect it, and so they may lack some of the intense color and "pop" of your monitor or HDTV. But they are portable, permanent, and easy to access. They bring back memories far better than any image on a screen.

And that's why we should bother with prints.

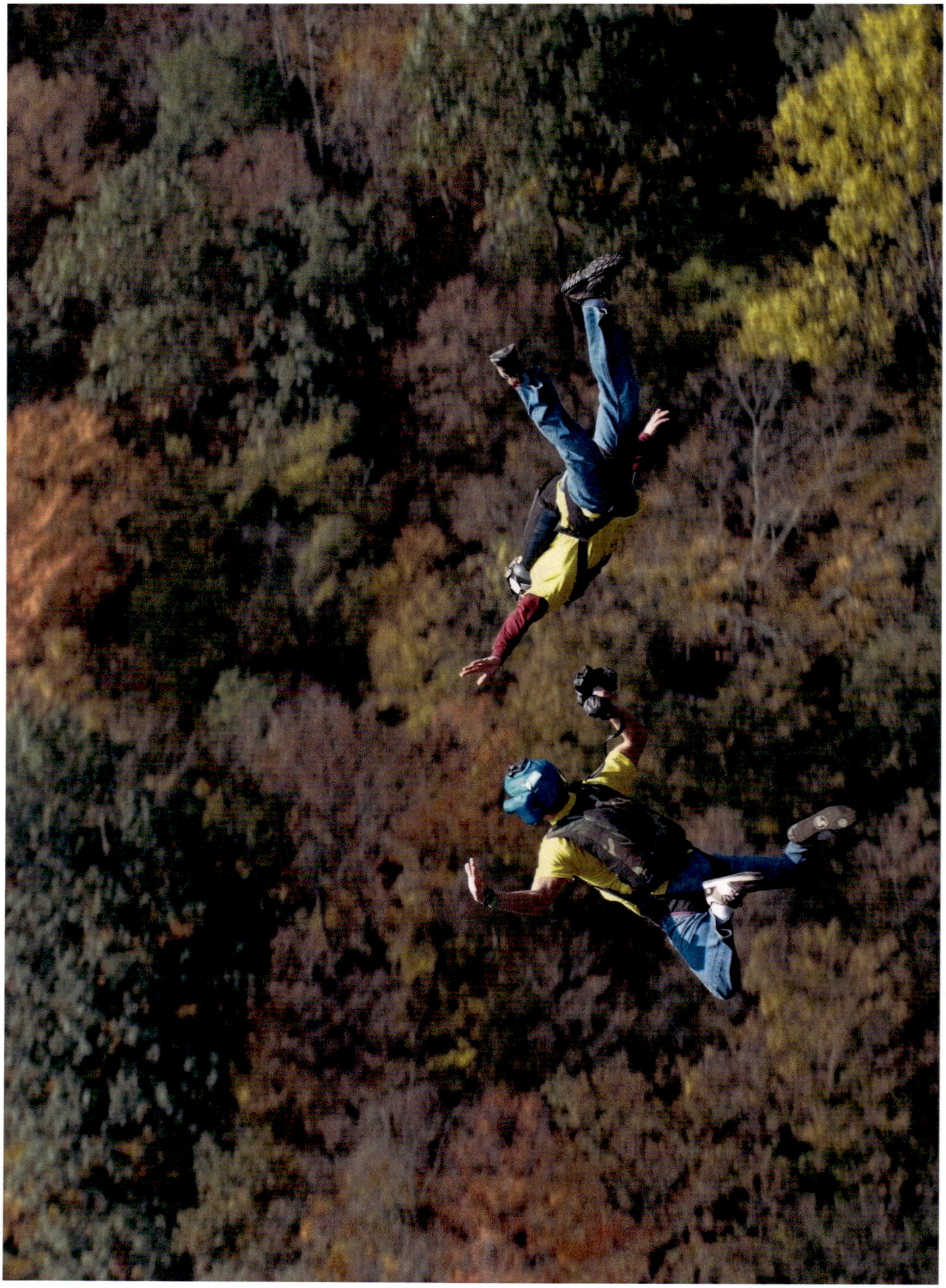

Figure 1-20 (left). *BASE jumpers at Bridge Day, Fayetteville WV*

THE PHILOSOPHY OF WOW!

We all know what a WOW! picture is. It's one that makes you say WOW! the instant you see it. Classic examples are the crash of the Hindenburg, the National Geographic picture of the Afghan girl with the intense blue eyes, the earth seen from the moon, the jetliner plunging into the World Trade Center, Ansel Adams' pictures of El Capitan or the lone oak tree in Yosemite... instantly recognizable, instantly effective.

The WOW! reaction is a sock in the gut. It's visceral, immediate, and intense. It's immediate in that there's an emotional, non-verbal reaction (other than maybe a spoken "wow!" or some such noise). The WOW! operates on the structural level-- no thought needed. The breath tightens, or gasps, the stomach may clench and relax, the pulse may race, adrenalin may surge, tears may flow -- a WOW! can do all those things. The emotion may be delight, or surprise, or horror, or love, or any of a cluster of feelings, but the most common is a recognition of beauty so intense you feel it both physically and emotionally -- even if the subject is horrific.

Part of what makes a WOW! is that it represents more than a selection of reality, or a flickering second of time. The shot is framed to exclude what's not relevant, and it catches what IS relevant at the precise moment when its meaning is greatest. But sometimes the WOW! comes from the IDEA or SYMBOL behind the picture, even more than from the image itself.

To get that WOW! lots of pictures are manipulated – lightened, darkened, cropped, recolored, blurred, sharpened, with details cloned out or in, vignetted, or (heaven help us) "HDR"-ed.

The design of a WOW! is one of the vital elements for success. Any photo orders reality; a WOW! just does this to a greater extent or a clearer purpose than other images. In a WOW! we don't see clutter, we don't see anything outside the frame, so there is a particular composition (arrangement of elements), color, tone, or object (recognizable or not) that evokes the WOW! reaction. WOWs are designed, intentionally composed -- even if the intention is a lightning-fast recognition by the photographer that THIS is the shot.

A WOW! may sometimes be a spontaneous "capture," but often they're thought through, manipulated on the computer or in the darkroom, printed with great care, and finally revealed to the viewer. Ansel Adams' creative process is meticulously documented, and he would spend up to three weeks in the darkroom working on a single image before he was satisfied that he had achieved the effect he had in mind before he ever picked up the camera.

Great photographers create a higher percentage of WOW!s than the rest of us, but only a tiny percentage of even the best photographer's work will be WOW!s.

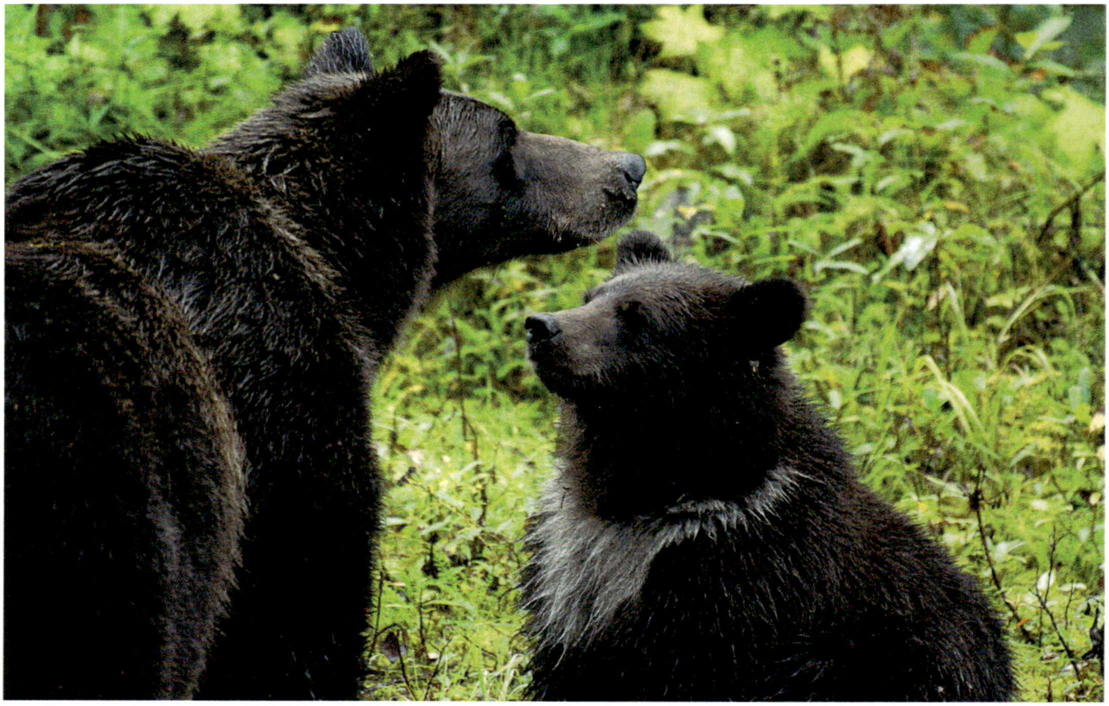

Figure 1-21. *Grizzly bear and cub, Hyder, AK*

Of course, what gets a WOW! from one person won't necessarily get a WOW! from everybody... we all filter what we see through our own lenses. But what's great about WOW! pictures is how broadly they DO reach people. A real WOW! grabs not one or two people, but thousands when they first encounter it.

Finally, I think the great power of a WOW! photograph is that it opens new possibilities for the viewer. It invites the viewer to react in new ways, to create new possibilities, to see the universe afresh. The WOW! is discontinuous, a sudden leap into another dimension. The gasp of sudden recognition that occurs so often on first glimpse of a WOW! is the opening, the rush of sudden insight and deep feeling.

STANDARDS FOR HIGH-END PRINTS, OR, A LESSON IN HUMILITY

If you've read this far, you've probably gathered that I'm fussy. Even finicky. Now mind you, I have friends who are far MORE nitpicky than I am --- they're the fussbudgets who tend to produce incredibly sharp pictures no bigger than 8 x 10, over which they've labored for the past two months --- you probably know people like them. THOSE folks use calipers to check that their french fries are consistent and fold their sandwich wrappers into origami. Now THAT's finicky. I'm not like that. Not at all. Nope. Never. No way! But yes, I'll admit to getting very anal trying to make an image right.

Now to the lesson in humility. I recently contacted a stock photo agency specializing in panoramas. This is not a microstock house; they are high-end and have been around for 30 years. Their instructions for submitting were detailed and precise: Adobe RGB, TIFF, 40 megs or bigger, sharp throughout the image, no dust spots, no visible stitch areas, etc. Submit 30-50 jpeg samples PLUS 5 full-bore tiffs.

No problem. These images were good! I sent in the required CD and waited. After a week or so I called to follow up, and the photo director of the stock house took the time to review some of the pictures with me. We looked at a number of pictures at 100% size. Not print size. Actual pixel size.

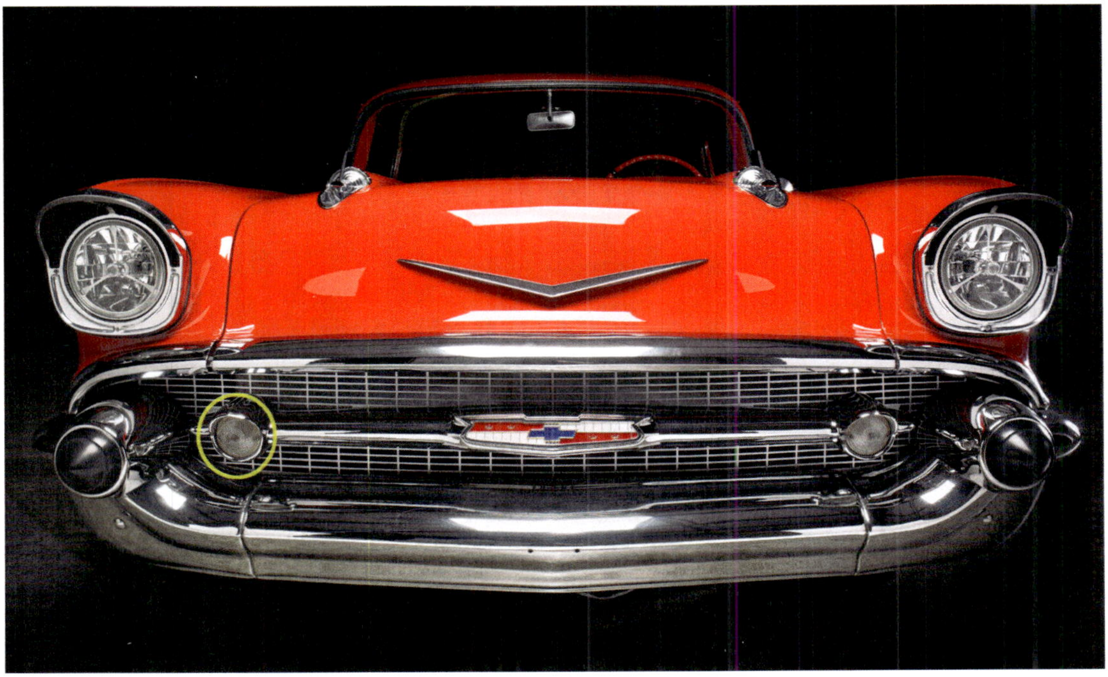

Figure 1-22. *1957 Chevy. Shot in studio. One that made it!*

"There, see, under the 22 inch mark, see, there's something off about the rigging on that trawler (1 of about 100 in the picture)."

"And in this one, see where there appears to be a fuzzy area, while the next bit is much clearer? I'm guessing that the first of your 3 or 4 images in this pano wasn't quite as sharp as the following ones. Won't do, I'm afraid."

A couple of my images were OK. As in two. Out of 30. I needed to resubmit additional pictures, now that I really understood the requirements. The Chevy was shot with a medium-format, 22 megapixel camera. Everything's there in this actual pixel size crop shown below, which is the circled area on the far left. It helped to have a couple of hours to light and take this shot in very controlled studio conditions.

This image is unsharpened, pretty much straight from the camera. This is the required standard set by the stock house. Here's a shot that didn't make the cut. Same camera, but depth of field too shallow. That's what I intended, but the image won't work for them.

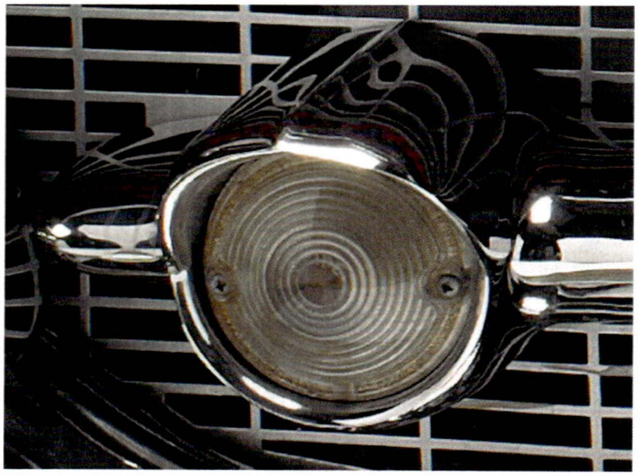

Figure 1-23. *Actual pixel crop of circled area in Figure 1-22*

As you may guess, this was a painful conversation. But it was VERY helpful, and I'm grateful to the photo director for taking the hour she spent with me on the phone.

I learned some things. First, if you're going to work with a camera under 10 megapixels, you have to produce a perfect print, drum-scan it, and hope, because you CANNOT res up an image to appropriate sizes without either some degree of pixelation or lack of sharpness. You're better off using a full-frame 12 megapixel or bigger camera, and medium format is better still.

I learned that sharpness counts ... to a degree you wouldn't believe. You can reduce noise and sharpen, but you have to do it VERY carefully to avoid even the hint of haloes or artifacts. And you'd better get rid of any chromatic aberration (purple fringing, for instance).

I learned that you really to have to be, literally, a pixel-peeper, as well as a darn good photographer. You have to review your images, and edit them, at actual pixel size. And in the panoramic world, you had better have that tripod level! Any kind of doubling or inconsistent sharpness – sayonara!

Now you can argue that NOBODY, but nobody, is going to print to a size where minor issues are going to show. And you'd be right. Unless the somebody is an art director who simply can't abide any kind of imperfection and who also has eyes sharp enough to see what you've missed.

But for rights-managed, high-end panoramas for stock, that's where the bar is set. As of this writing, I'm waiting to learn if the additional images I submitted do pass muster and I can go under contract to these folks. I hope so. But I really appreciate learning a little humility and recognizing that my work can certainly take another step up technically. Now if I could only afford a new Phase Camera with 75 MP back, the technical side would be no problem. But failing that, I'd better emulate my truly anal friends and pixel-peep with the best of them.

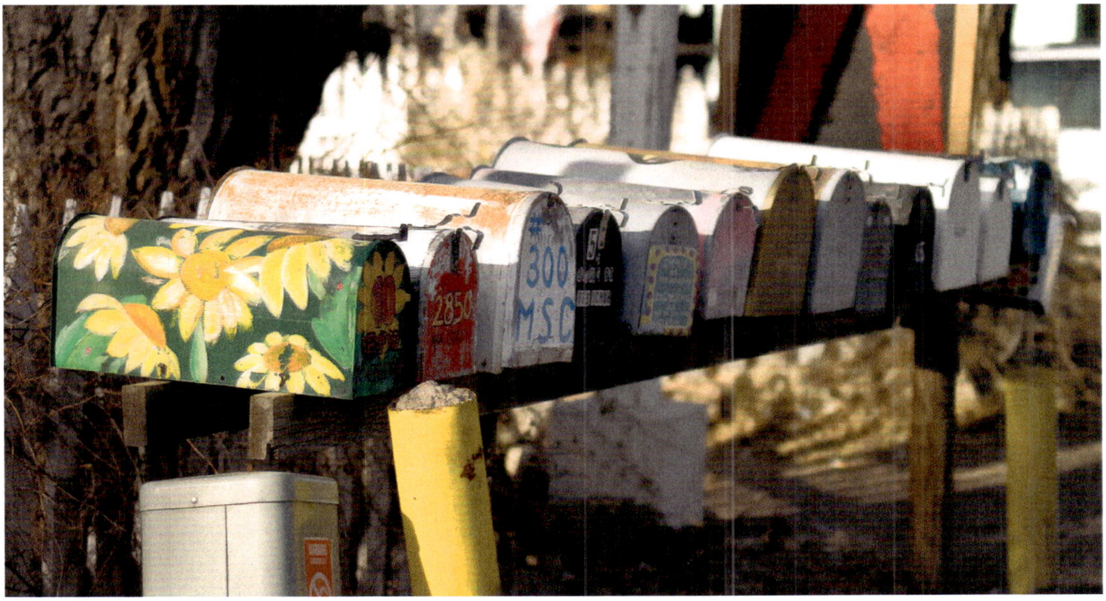

Figure 1-24. *Mail boxes in Madrid, NM.*

CHAPTER 2: *Attitude and Motivation*

KEEPING THAT SENSE OF PLAY
OMG, THAT'S MY BABY (EFFECTIVE VS. INEFFECTIVE CRITIQUE)
THE RISKY BUSINESS OF ELIMINATING RISK
DON'T PRETEND YOUR BEST WORK HAS VANISHED

CHAPTER 2: *Attitude and Motivation*

KEEPING THAT SENSE OF PLAY

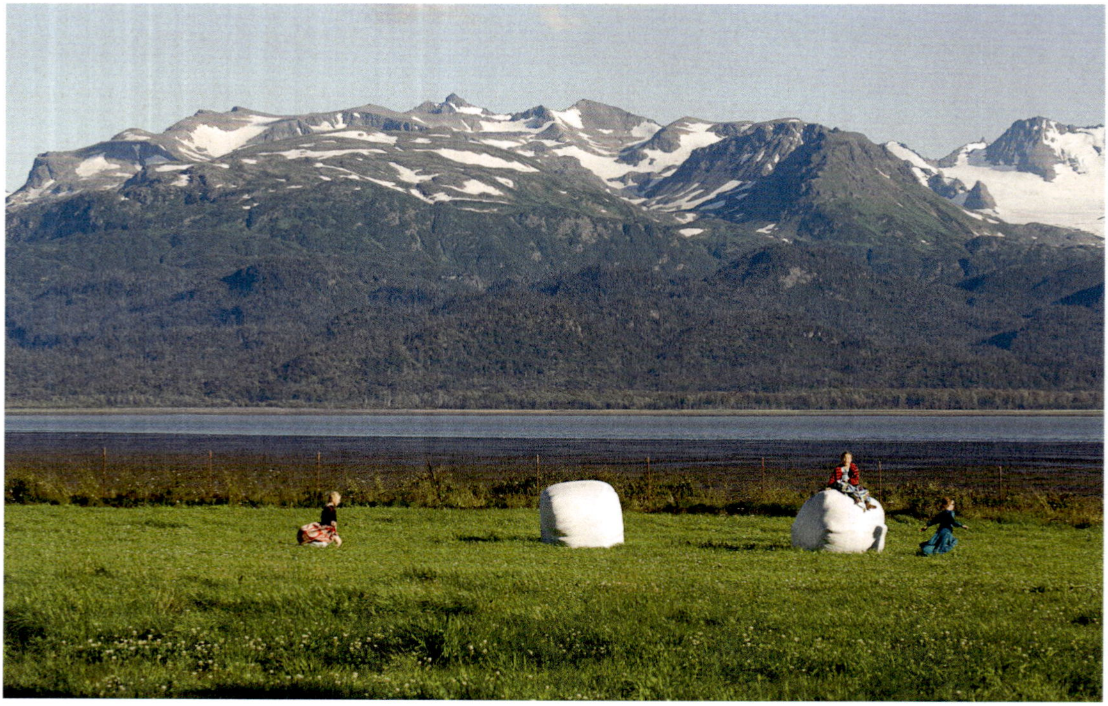

Figure 2-1. *Kids playing on hay bales in Kachemak Silo (Old Believer's Village) near Homer, AK.*

One Saturday my friend and web designer Tom Seibert stopped by the studio. Like many of my "amateur" friends, Tom is a better photographer than he thinks. Together we looked at a number of his pictures and at a number of other people's works that he admired.

I found myself profoundly jealous. Because Tom doesn't know "the rules," and doesn't have to follow them to earn his living, they don't bind him. He brings a perpetual sense of play to his photography, and both he and the work are the better for that.

By play, I mean he's willing to try stuff just to see what happens, without a fixed outcome or "salable print" in mind. He's just following his passion (as time permits), making things up as he goes along, and having a wonderful time doing it. His eye and his imagination are working together, and the result is often terrific.

You can visit TomSeibertPhotography.com to get a sense of the guy and his work.

What happens to many of us when we turn pro is that we lose that sense of discovery and play that Tom (and many other amateurs) bring to photography, though that doesn't seem to have happened to Tom. Turning pro, we have to achieve predictably good results that will please the customer, time after time, and in as little time as possible. Time, after all, is money.

We learn that innovation is fine— in certain channels. We learn about the "PPA corner" and other approved ways of composing photos. We learn that cute kids and smiling moms outsell sad kids and pensive moms 50 to 1 – so we don't even try it the other way.

Not all professional photographers fall into this trap. The commercial and product photographers in particular get to bring something extra to the job. And whatever the specialty, some photographers hang on to their artistic playfulness and willingness to just try something new.

I'm not suggesting in this little essay that photographers should willfully ignore technique, and just "chimp" away. That's not what play is about. Play is "what happens if I try shooting this shot from ground level instead of standing up?" But if you want to grow as a photographer, you take notes (even if only mental notes) about what you did and the results you got.

I'm sure that somewhere in the distant past a caveman got curious about the shadows his hand was casting on the walls – he fooled around with it and invented shadow puppets. After that, perhaps he learned how to make a number of different animal shapes. And maybe traced some of them on the wall of the cave with charcoal, just to see how it worked out .... and art was born. Or maybe it was religion. Or maybe he just entertained the kids in the tribe with his nifty trick.

What matters is that play can lead to interesting, valuable, and repeatable results --- sometimes. Being willing to accept the goofs, mistakes, and ugly results is what makes this kind of exploratory play authentic – and fun.

"Amateur" doesn't mean incompetent. It means someone who loves what they're doing. The difference between amateur and professional has more to do with the expectations of earning a living than it does with quality. And if we professionals lose our ability to play, we are losing something more important than money. We're losing a lot of our chances to grow and invent our art anew.

OMG, THAT'S MY BABY (EFFECTIVE VS. INEFFECTIVE CRITIQUE)

Some years ago, when I was making my living as a writer and editor, a lot of people would come to me and say something like "I've written this book. Would you take a look and tell me what you think?" It quickly became obvious that NOBODY wanted to know what I thought. What they wanted was for me, as an official expert, to bless them and their work.

Photographers have exactly the same need for reassurance, not critique. And that's a pity, because effective critique can really help photographers do a better job.

But here's a home truth: when a photograph is completed and released to the world, whether via Flickr, a photo competition, or even by cell phone, we no longer own it (legally, yes, but that's a different topic). It exists independently of the maker and will be welcomed or condemned by whoever sees it.

Unfortunately, most of the time photographers, like writers, don't see that their baby is now grown up. As long as they see the photo as a part of themselves, even constructive critique will be hard to take. And it's too bad, but there's far more bad critique than helpful critique, and this blog entry is intended to make the difference clear.

Ineffective Critique vs. Useful Critique

Critique is ineffective when it fails to give the photographer specific, concrete advice based upon commonly accepted quality standards.

The least effective critique is "I don't like it" (or the reverse! "I like it" doesn't give anything to help the maker do better).

The key thing missing is "WHY" I like it or don't like it. And the "why" needs to be put carefully. It needs to be concrete, specific, refer to the artistic principle or technique involved. Above all it needs to tell the maker something he or she can do to improve the image.

Here are some examples of bad critique, along with improved versions that are more constructive.

Bad: This picture of an old car sucks. I hate it!

Better: This picture of an old car is confusing; I can't tell what it's about. And it's too dark.

Best: There's no center of interest in this picture of an old car. The flag is distracting, and the dark shadows in the engine compartment rob that area of meaning. The headlight on the bottom pulls your eye because it's so bright, but the picture isn't about the headlight. Re-crop, lighten the shadows, and it will be much better.

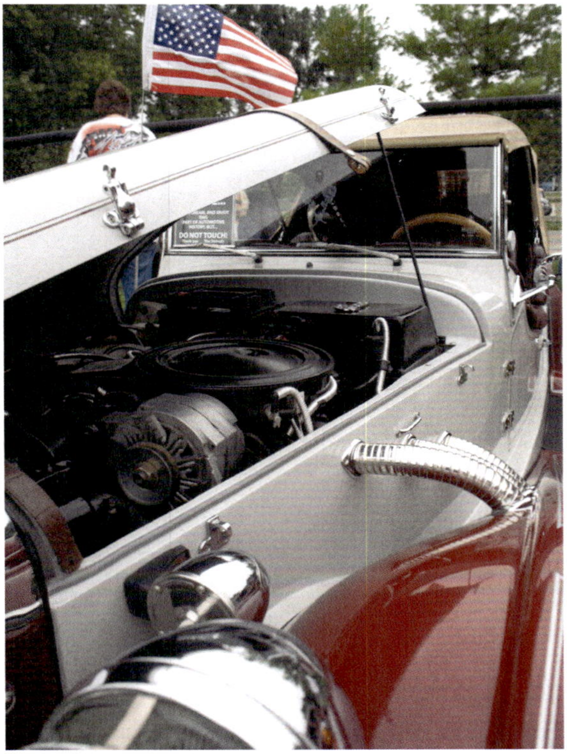

Figure 2-3. *Vintage Car show, South Lebanon OH*

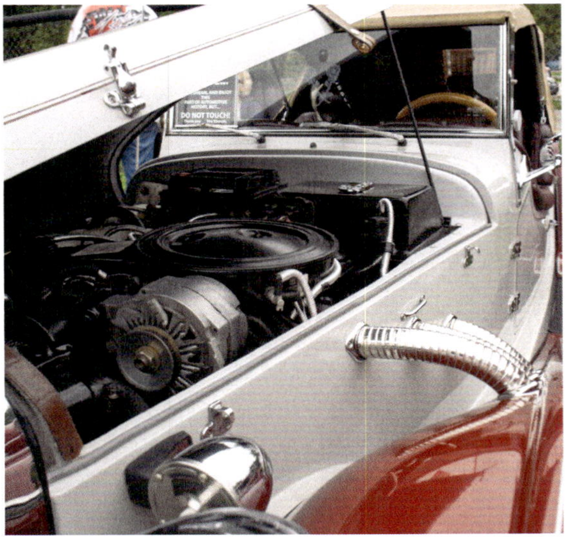

Figure 2-4. *Improved image*

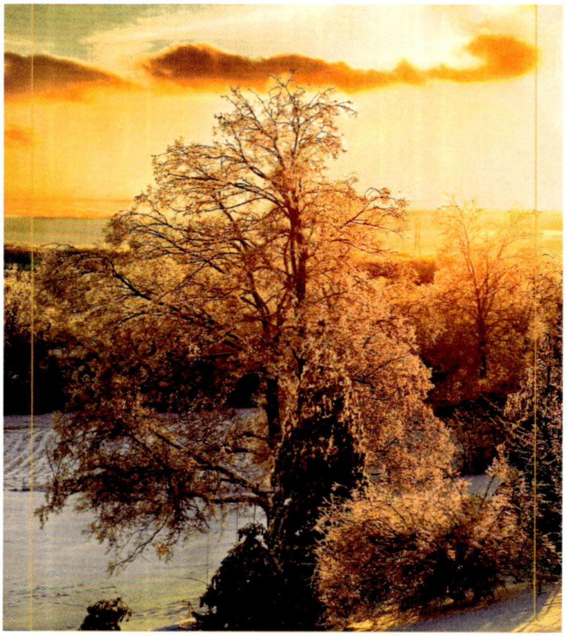

Figure 2-5*. Storm Sunset, Loveland, OH*

Bad: Bad composition!

Better: No central focus, and there's a dark area which creates a hole in the scene.

Best: Turn the picture upside down and you'll see that there isn't a central subject of interest and that there's a hole pulling the eye away. Recrop to eliminate the hole and emphasize the icy tree. And remove the distracting power tower while you're at it.

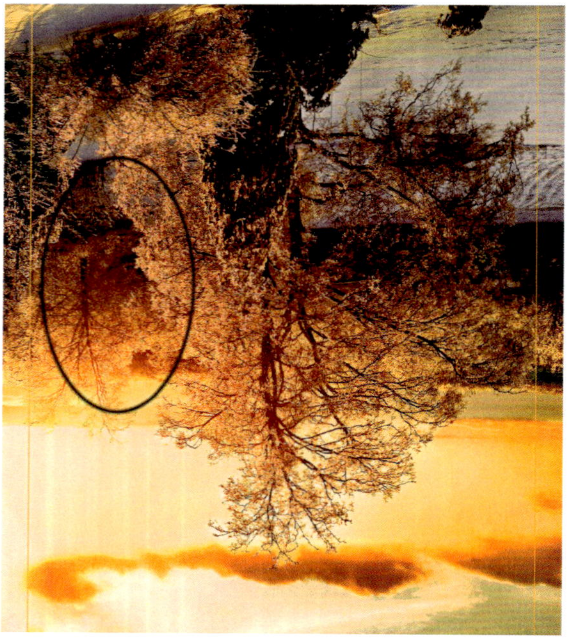

Figure 2-6*. inverted version of 2-5, ice storm sunset*

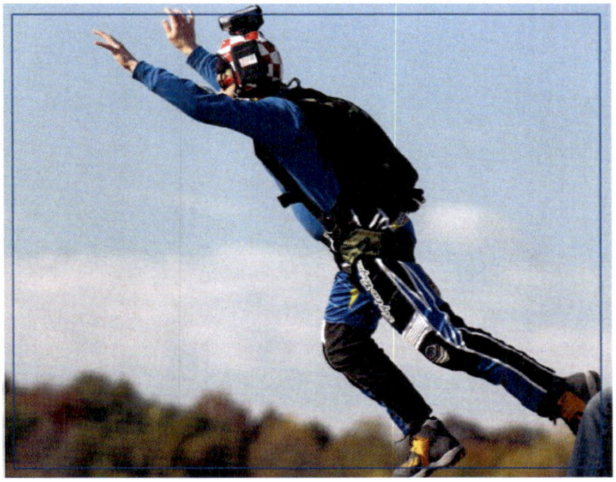

Figure 2-7. *BASE jumper, New River Bridge, Fayetteville, VA*

Bad: The blue jeans and knee in the lower right corner are distracting. Shoot it again without them.

Better: There is no better for this one. It's obvious that this is an event photo and there is no way to get a clearer field of view. When there's nothing to be done to improve the shot, the critique is actually offensive. Sometimes there's just nothing you can do! At the same time, while you can't move mountains or boulders, you CAN decide this shot's either worthwhile despite the problem because it tells a compelling story or has a lot of impact (Wiley Coyote, look out!), or you can decide not to take it in the first place --- or keep it for your private pleasure.

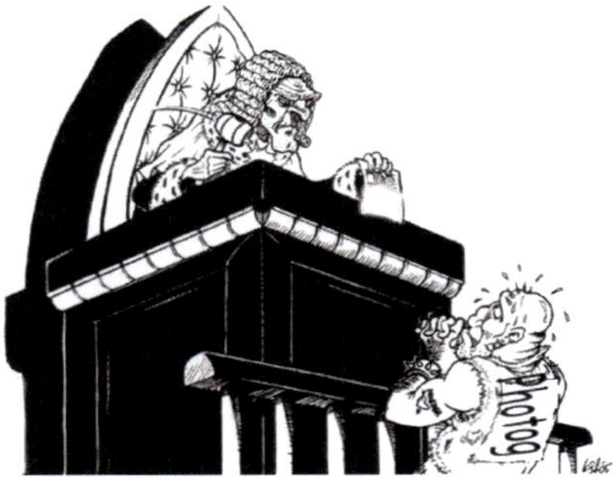

Figure 2-8. *Have mercy on us, Judge!*

So, to sum up – critique scares most photographers silly. As a receiver of critique, you have to be able to say it's the work, not yourself, that's being critiqued. (True, because you no longer "own" the work). On the critic's side, try to give critique which is practical, do-able (unlike moving mountains), and based on sound principles. One-word judgments (bad composition, bad presentation, no subject) are not helpful! Finally, critique can be the artist's best friend or most dreaded enemy. It takes two to make a good critique!

THE RISKY BUSINESS OF ELIMINATING RISK

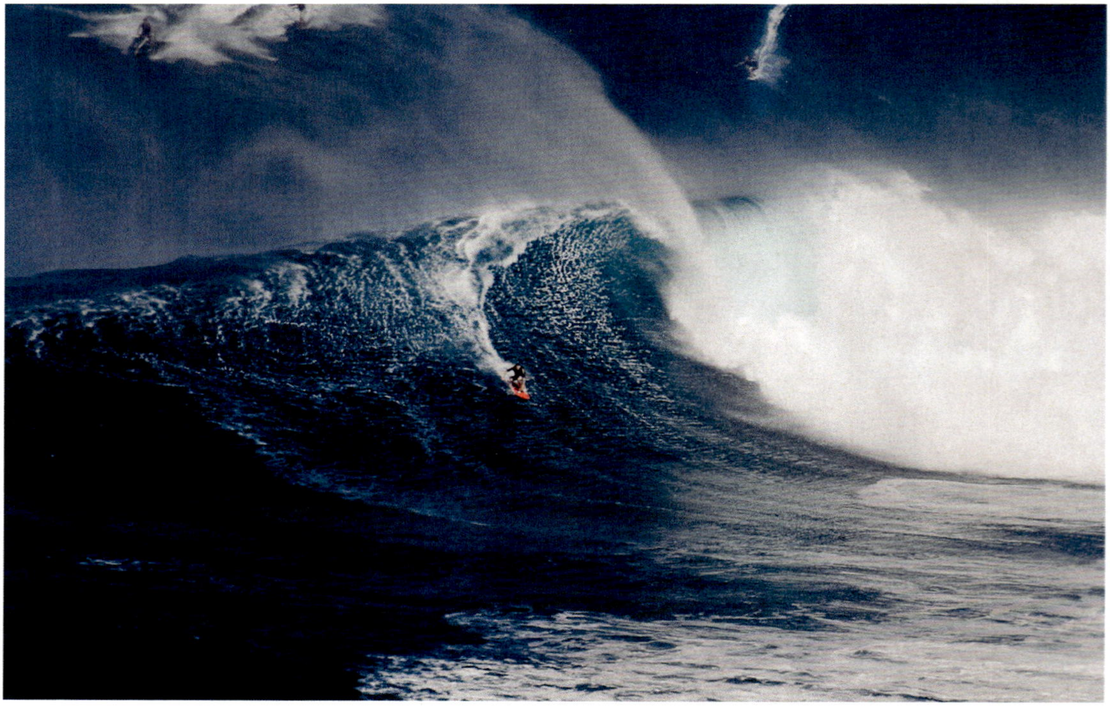

Figure 2-9. *Monster Surf on Maui. You cannot eliminate risk in every situation!*

As I was pounding the treadmill uphill to nowhere the other day, CNN was covering a flash flood in Arkansas that swept away scores of campers, with 20 reported dead. The anchor asked in all seriousness "how can we prevent a repetition of such things in the future?" She really believed that government has to DO SOMETHING when random events occur and people are in the way. Worse, she appeared to believe government COULD do something to make camping safer in the event of other freaks of nature, or forbid it altogether.

This is flat-out idiocy.

Believing you can sanitize the risk out of life is arrogant. Crap does happen, and you can't always protect against it or predict it. If you think it's possible to achieve 100% safety, you are 100% wrong, and you will eventually be 100% dead. That's the way it goes.

As photographers, we face all sorts of risk. Some of what we do is flat-out dangerous. Some exposes us to various forms of liability. And we are always exposed to the risk of people not appreciating our work.
If risk cannot be eliminated, it can be managed.

There are a few key questions to ask that can help put risk into perspective. These originated with Kepner-Trego in their work *The New Rational Manager.*

1. What can go wrong?
2. What are the consequences if it does go wrong, and how serious are they?
3. How likely is it to go wrong?
4. Is it worth the risk?
5. What can I do to reduce the risk?

If this sounds like common sense, that's because "common sense" consists of considering one's actions in the light of these questions, whether consciously or unconsciously. Failing to consider them at all is foolhardy.

In the case of the campground in Arkansas, clearly flash floods are on the "what can go wrong" list. The severity of the consequences are hard to imagine BECAUSE the frequency / likelihood of flash floods is so low. You can reduce the risk by camping elsewhere or by tethering yourself to a tree, or by finding higher ground BEFORE pitching your tent, or by listening to the weather forecast before deciding to camp there. But if you are going to camp there, it's a risk, however small, that you agree to take.

Where does this notion come from, that complete safety is both possible and a personal right? That risk, any risk at all, is some moral villainy to be hated and feared?

The pioneers who arrived in Virginia and Massachusetts ran risks that are unthinkable today. And as western culture spread across the continent, the grit it took to face enormous risks travelled with us. Avalanche, starvation, disease, hostile natives, just plain breaking a leg and eventually dying either of gangrene or of thirst because you can't get to a water supply —risks like these were the normal fabric of life from 1620 til well into this century.

These huge risks killed off lots of people, but they didn't deter the pioneers from moving west to farm new lands, or prospectors from flooding into the Klondike around 1900.

We thought nothing of taking risks—and we didn't blame government every time we lost our bet with fate, or when we did something truly stupid and suffered the consequences. Our forebears had true grit. They couldn't have done what they did without it.

Flash forward. Today every mountain path is supposed to have stairs and handrails, right? Every campground needs a central alarm, and someone to monitor it 24/7 (as was proposed in the aftermath of the Arkansas flash flood) or individual squawk boxes you carry with you to your campsite. We want to see bears, but we expect them to be tame and predictable and cute.

Bah!

Now how does all this apply to us as photographers? First, there is physical risk. If you want the shot that nobody else has, you have to be prepared to go places other people don't go. You have to climb slippery rocks, descend steep slopes, and hang out over precipices. If you want this waterfall shot, here's where you have to stand:

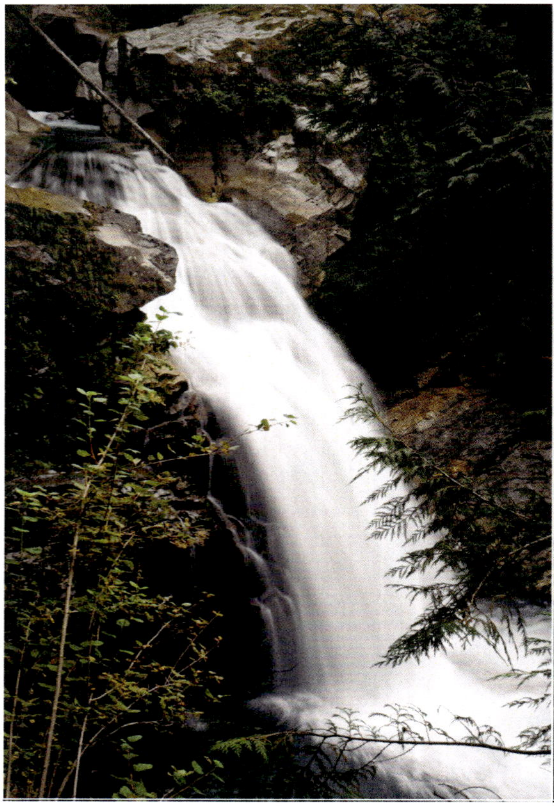

Figure 2-10. *Washington State, 2009.*

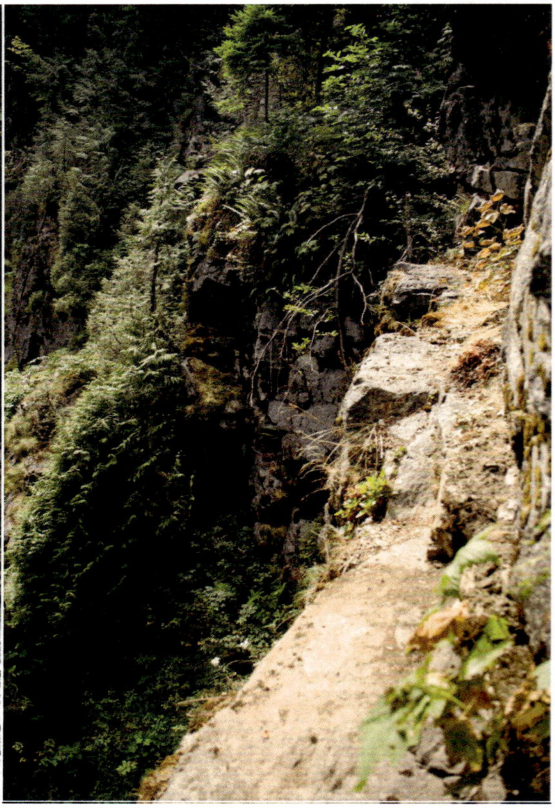

Figure 2-11. *Washington State, 2009.*

That's just the way it is. This is a risky photograph. To go back to the 5 questions, the seriousness of a mishap is very high, and the likelihood moderate to high (these are subjective values; if you're a natural-born-mountain goat with young legs, both the risk and likelihood are lower). So the first question to ask is, do you want the shot enough to run the risk? How can you lower the risk so as to make it safer, for instance, by roping yourself to a strongpoint somewhere, changing your shoes, leaving your camera pack back on safer ground and taking only yourself and your tripod out there? To take the surfing picture (Figure 2-9 above,) I had to work through the tall grasses til I got to the edge of a 100 foot cliff, and sit on the edge with my legs dangling, along with two Hawaiian photogs. There was room for 2.5 of us, so the crowding didn't help — but I knew if one of us started to go, the others would at least save the gear!

It's not only nature photographers who run risks. One of these days some 8-year-old terror is going to run through your camera room, hit a lamp, which will fall (satisfyingly for him) on his little sister, sending her to the hospital or worse— not to mention ruining your lamp.

This is another real risk, perhaps less likely, but potentially very, very serious —risk management here is to properly secure your lights as best you can, warn kids not to stray, and carry good liability insurance. Someday someone will sue you, and that's a guarantee in modern America.

Finally, there is also emotional risk – see the article on Critique elsewhere in this chapter. If you want to grow as a photographer you have to run the risk of rejection. It's a real risk, and it's painful, but it doesn't kill you. You can manage that risk by realizing it isn't personal — once it's out there, the photo isn't about the photographer, it's about the photograph itself. You can also manage the risk by looking at what you gain vs. what you risk, and just decide to accept the risk, knowing it may not work out the way you wanted. Of course, the best way to manage the risk of publishing your work is simply to do very good work in the first place! But even then, there's no certainty folks will like it.

I place financial risk under the category of emotional risk. The odds of succeeding as a fine art or nature photographer are ridiculously long. The only way to manage this risk is to realize that marketing is even more important than the quality of your photographs ... and most photographers detest the marketing process. Maybe you're better off staying in the amateur world.

So, to conclude: risk is everywhere in life, and outdoor photographers face risk more than many other folks. You can manage risk, and if you do, and you fail, at least you fail forward.

DON'T PRETEND YOUR BEST WORK HAS VANISHED

The blog Permission to Suck suggested that the way to creative freedom is to pretend that all your work has vanished. All of it. Here's what the blog said:

> Just for a moment act as though all the creative works you've ever done are gone and untraceable. They cease to be. There is no memory of your body of work except for what you own; friends, family, colleagues, students, peers – all have no recollection of your work or your reputation. There is no longer a creative history to foster your pride.
>
> Assuming your passion is not make-believe and your ability to produce remains, now what? Your beautifully clever, well-crafted, richly toned works are disposable. a.k.a – You're free.
>
> It's important that we know our best work is yet to be produced. Our best is what is in us now, not what we – or others – admire about the past. The current creative paradox or those conflicts that remain unresolved are in front of us and are what needs examining

I understand that the attempt here is to free oneself from others' expectations. That's a fine goal. But if you write (or photograph or paint) ONLY for yourself, taking no heed of the world outside you, you become divorced from human contact and take no responsibility for how your actions (or your work) affect others. In other words, you risk becoming a kee-kee bird, which flies in ever decreasing spirals until it flies up its own butt and cries out "kee, kee, kee-rimeny it's dark in here!"

One line of thought, very hard to accept (which doesn't invalidate it), is that the ONLY way we have an identity is through the collective responses of our community to us. Without a community, we vanish, goes this argument. I have a tough time buying that. My own belief is that you should shoot / draw/ write for yourself FIRST, but understand that until the work is out there in the world it has no effective existence. ("Effective" means it brings things about or causes things.)

I do a lot of photo restoration. It pays the bills, and it REALLY registers with customers. In essence, and to their way of seeing, I am literally bringing loved ones back to life. That is what defines the work as effective: the belief of others that it has validity and worth. The goal of restoration is to bring the customer's memories back to life. Artistic perfection is nice, but it is not the point. The point is how it makes the customer feel. But artistic validity and worth in the mind of the herd may not be the same as the validity and worth provided by sophisticated viewers / readers.

Basically, many writers and photographers are really insecure about their stuff, and they DO try too hard to please others. It doesn't follow from that that they should pretend no one out there exists.... what if everyone out there pretended YOU don't exist? It has to be a balancing act between your own creative urges and the perceptions of your audience.

This entire problem is made more complicated by the lack of control we as artists face when we send our goods (read, a portion of ourselves) out there for others to judge. Scary! It's why I no longer review the work of beginners, because I know they don't want critique, they want reassurance ... which in conscience I may or may not be able to give them.

Once your work leaves the nest, it's not your bird any longer. And it definitely is not you. The minute you publish a work of art, or a blog entry, or this book, it no longer belongs to you (except in a legal sense, and good luck enforcing THAT these days). It is an independent object, just as a toenail clipping is when it leaves your foot. Whether it shines or it stinks, it's no longer yours to command. It's out there in the universe, and the universe will judge it by word or action.

As artists we tend to put so much of ourselves into our work that it is a tremendous strain to let the stuff go. But we have to, or we're kee-kee birds.

So I say DON'T pretend all your stuff has vanished. Keep your best stuff, and when you work up the nerve, produce even better work. And when you really believe, really, really believe your stuff is great, send it out into the void and accept the consequences. If it's great enough, it will light a fire somewhere!

CHAPTER 3: *Travel and Photography*

CHAPTER 3: *Travel and Photography*

THE MAGICAL ACT OF ASKING YOUR WAY

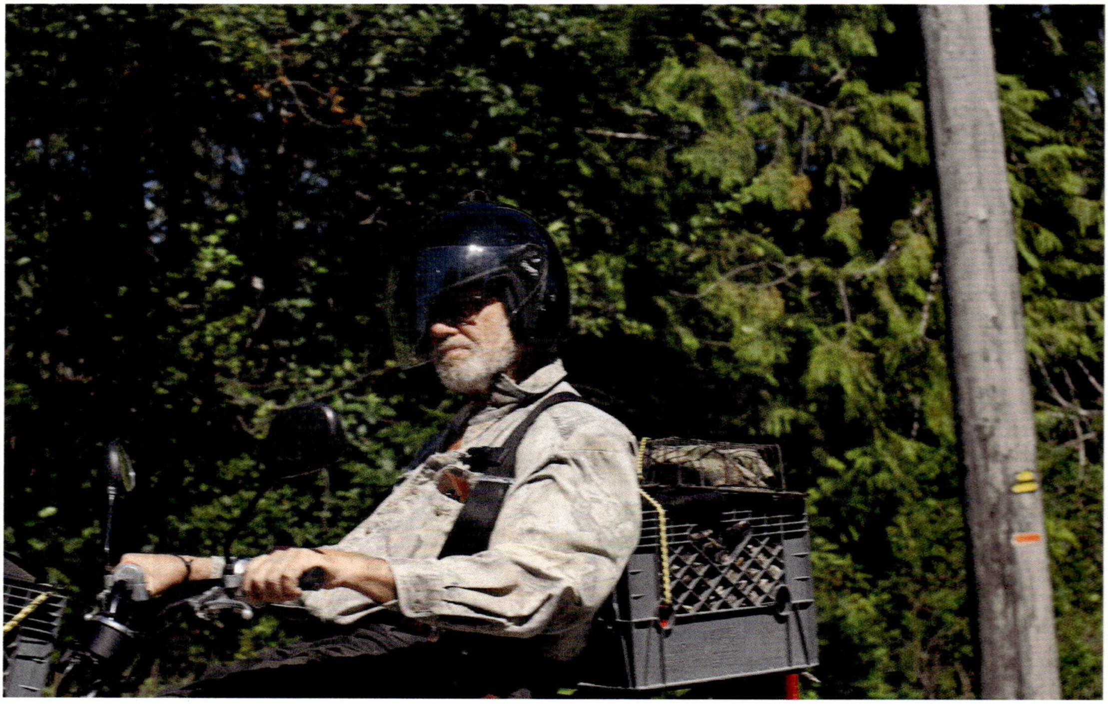

Figure 3-1. *Washington Local. He knows where to find anything!*

"Why does it take 20,000,000 sperms to fertilize a single human egg?"
Because not one of them will stop to ask directions!

Yesterday I was on a waterfall hunting expedition with three friends, Tina, Tom, and Mark. Tom was driving. We were trying to find Deadman's Falls near Georgetown, OH. We cruised up and down route 132 hunting for what we thought was the right road that would lead to the field we'd have to cross. Finally, we turned into a road with a sign for a winery. We needed to ask for directions – yet even when we saw a fellow standing out by his mailbox, Tom had to be bludgeoned into turning around and asking him the way.

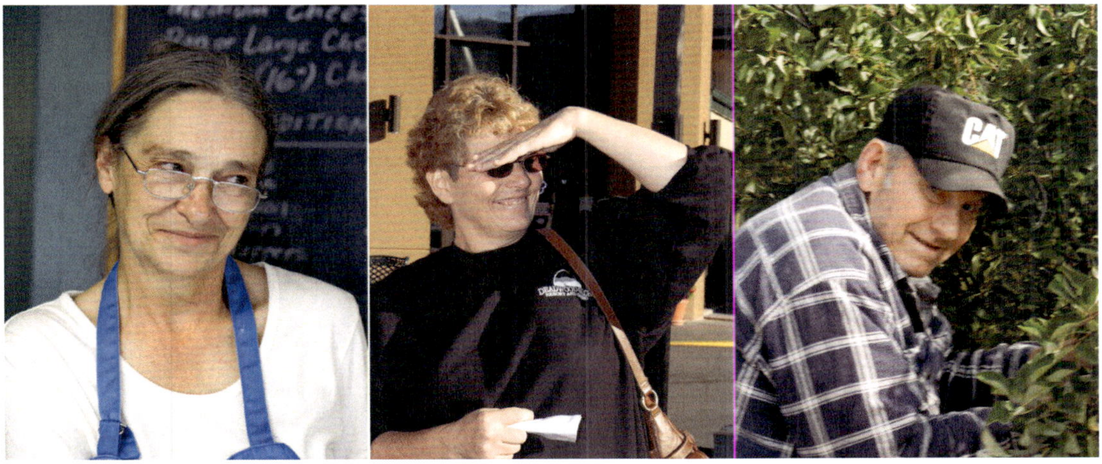

Figures 3-2, 3-3 and 3-4. *People are glad to provide help if you ask them.*

The fellow turned out to be a very nice and friendly farmer, and he gave us good directions. Definitely a twinkle in his eye!

This small incident prompted a conversation and some reflection about the difficulty so many of us – especially men – seem to have in asking for directions. The little joke at the top of the page wouldn't be funny if it weren't a truth universally acknowledge among women that a man – whether or not possessed of a fortune – really hates to ask directions. But it isn't just men, though there may be a bias towards men.

Somewhere out there must be a PhD dissertation on this topic, but I haven't seen it. So I'll wonder. Are men more reluctant to ask other men for directions? Do women cozy up to other women but shun the gents when it comes to asking the way? Or is it the other way round? Is this reluctance to ask cultural, geographical, age-related, or genetic? Or is it simply a reluctance to admit one's vulnerability in seeking help from strangers?

My guess is that the last explanation is most at play and applies more to men than to women but doesn't exclude either gender. When we stop someone to ask them the way, we may tell ourselves lots of stories:

1. There's something wrong with me that I can't find the way, and I'll look bad if is ask.
2. It's bad manners to invade someone else's space, and I don't want to be rude.
3. What if this guy is hostile? He could get mad at me for bothering him!
4. What if this guy's hostile, and shines me on with WRONG directions, just to be mean?
5. What if he thinks I'm trying to sell him something?
6. Maybe there are dogs in the back yard!
7. This place is run down. What if there's a meth lab in the back yard?
   (Not unheard of in rural Southwest Ohio!)

Of course, if it's New York City, New Jersey, Kabul or Cartagena, any of these stories might be true — even all of them at the same time! But most likely, even in those places, none of them is true, or only rarely true.

If you stop to examine your feelings (not a very guy thing to do), you'll probably see you have some connection to these stories. Or perhaps you have one all your own that blocks you from asking for directions.

In fact – and here's a secret for travel – most places, people are DELIGHTED to be asked for help, and will offer it gladly when you do. I've had incredible offers of help in France, Alaska, Canada, Mexico, Nova Scotia, Arkansas, and – yes – Manhattan (New York, not Kansas). And I STILL don't like asking directions, but I do it – and am usually rewarded for it.

You see, when you ask for directions (or other help), you are making yourself vulnerable, opening yourself wide open as a flawed human being who doesn't know absolutely everything. When you do that, you instantly cease to be a threat or challenge to the person you've approached. Instead, you awake feelings of generosity that can run to amazing lengths – such as the person who not only gave me directions to an out-of the-way site, but also led me there. Or the old man in France who actually took a traveler's check from my 17-year old self so that I could get something to eat – it was a religious holiday and no banks or stores in the village were open.

Put yourself in the position of being the one asked for help. What feelings race through you? And if you provide the help, how does that make you feel? (The answer, just in case you've never actually helped a stranger, is YOU FEEL TERRIFIC.)

Sometimes an act as simple as asking (or giving) directions can lead to genuine friendships that can last for years and years. You may wind up meeting new people and learning lots that you didn't know about the place you're in, the things you might never have known.

On my two-month trip to Alaska, I made a conscious decision to talk to strangers. Now up there, that's not so hard to do, because eye contact is not a forgotten art as it is in more densely packed urban centers. And in these random conversations, I'd always ask one question – "You've been a terrific help. Now on my travels, who else should I talk to?" Amazing! That simple question connected me (or at least gave me entrée) over hundreds of miles and wound up providing riches untold.

Asking directions is a conscious decision. You have to actually choose to do it.

The trick in freeing yourself up to ask directions is being aware of how silly are most of the things that hold you back. It's easier if you actually ask yourself, "How true are these stories I've been telling myself? What's the worst thing that could happen, and how likely is it to occur?"

We may not like asking strangers for help, even just asking for directions. But in the act of asking for help, we are actually doing the stranger a great service – we are validating them for their possible kindness, their knowledge, and their experience. What a gift to offer to anyone – and how rarely will it be turned down.

That is the great secret and marvelous blessing of asking your way – it's really a two-way street. So, stop the car, get out of it, and approach that guy leaning on the mailbox. It's an act of kindness and a true adventure!

FINDING HIDDEN LOCATIONS

What photographer doesn't dream of finding a secret beauty spot, something no human eyes have seen? Ain't gonna happen, bubba. With the exception of certain inhospitable places like Antarctica or the center of equatorial Africa, there just aren't any unseen-by-humans spots left. There are, however, plenty of rarely visited spots, rarely used vantage points, and rarely photographed sights — if you know how to find them.

Finding such spots comes down to five things: research, inquisitiveness, effort, perseverance, and luck.

Since the arrival of Google Earth and smart phones, research has gotten easier and easier. But do you know how to recognize a waterfall from a somewhat murky picture from 3,000 feet?

Here's an example. Could you make the leap from Google Earth to reaching the falls shown in the second photo (marked by a yellow pin)? Can you find a second falls (unmarked) in the same Google Earth picture?

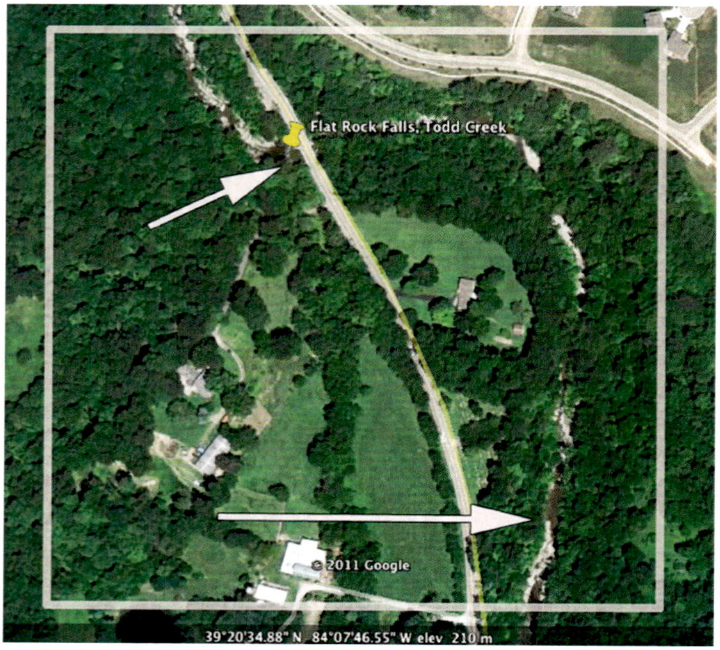

Figure 3-5. *Google Earth. Yellow pin marks Flat Rock Falls. Lower arrow indicates second falls (upstream from Flat Rock). Using these views, like using topo maps, takes practice.*

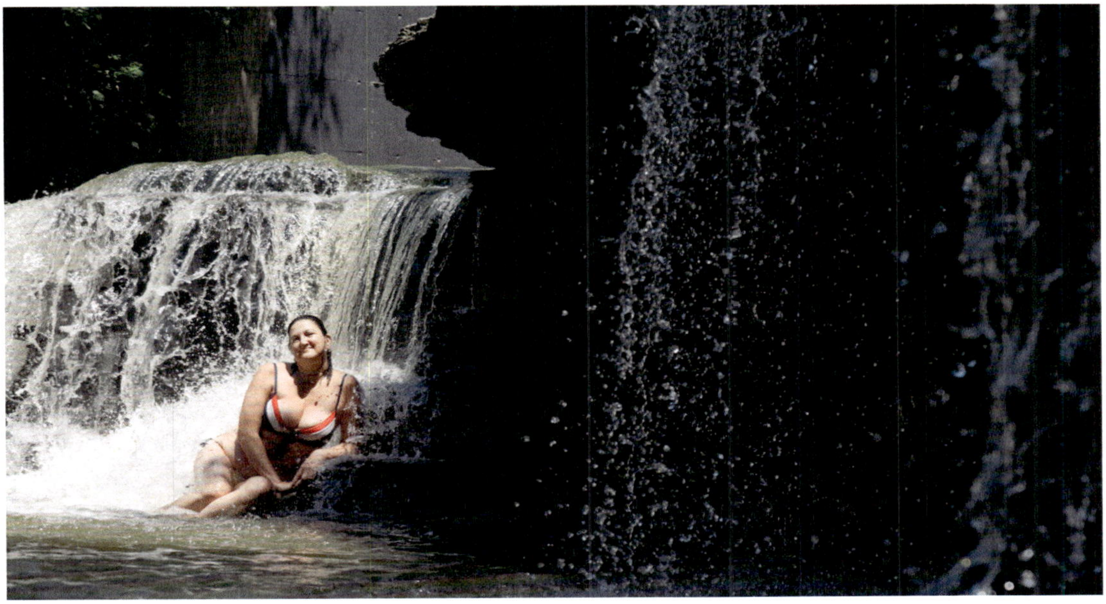

Figure 3-6. *Flat Rock Falls shown on Google Earth in Figure 3-5*

The point is that you need to practice interpreting Google Earth and other mapping software to find promising locations for a shoot. Some folks are better at this than others (and I'm one of the others!), but practice does help.

Likewise, you need to learn how to use topographical maps. Again, picking on waterfalls for a moment, look for streams with "pinch points" where narrow cliffs run together, especially when accompanied by sudden flattening out. But using topographical maps DOES take practice, and if you misread them you can get into quite a bunch of trouble.

Don't forget guidebooks. They may mention places that really are off the beaten tourist track – but expect to put some real effort in getting to them.

Historical literature and travel literature can also guide you. For instance, I learned about another small falls near my studio. The studio's in an historic building, and I was digging into its past. I ran across a reference to a small, seasonal sawmill located on Bear Run. Hmm. Never heard of it. But seasonal mills = waterfalls, and so I found them. (They're on private land, so I can't tell you how to get there).

Research, using maps, newspapers, and mapping software can give you a good lead on little-known locations. Hiking clubs trail guides, and kayaking clubs also print guides to their areas of interest. Unknown? No. Not-so-well known? You bet.

And here's the source that's worth its weight in gold: ASK A NATIVE. (How to get to know natives? Read "Eric's Rules for Successful Travel.") In Dayton, Wyoming, I saw a sign that intrigued me, so I took the long gravel driveway to arrive at a monastery-cum-retreat-center. I wandered around a while and encountered a vivacious woman of a certain age who introduced herself as the head of the order. I blinked a few times, because she was obviously female, and the place was clearly a monastery, but hey, it's not 1340 any more. After we talked awhile, she suggested a hike along the Tongue River Canyon, and warned me to avoid the fork and bridge that would lead back into a cave system. (Now THAT would have been a great photo site). But I followed her advice to get to a location, invisible from any road, and well worth the hike. That's the "eye of the needle" formation seen below.

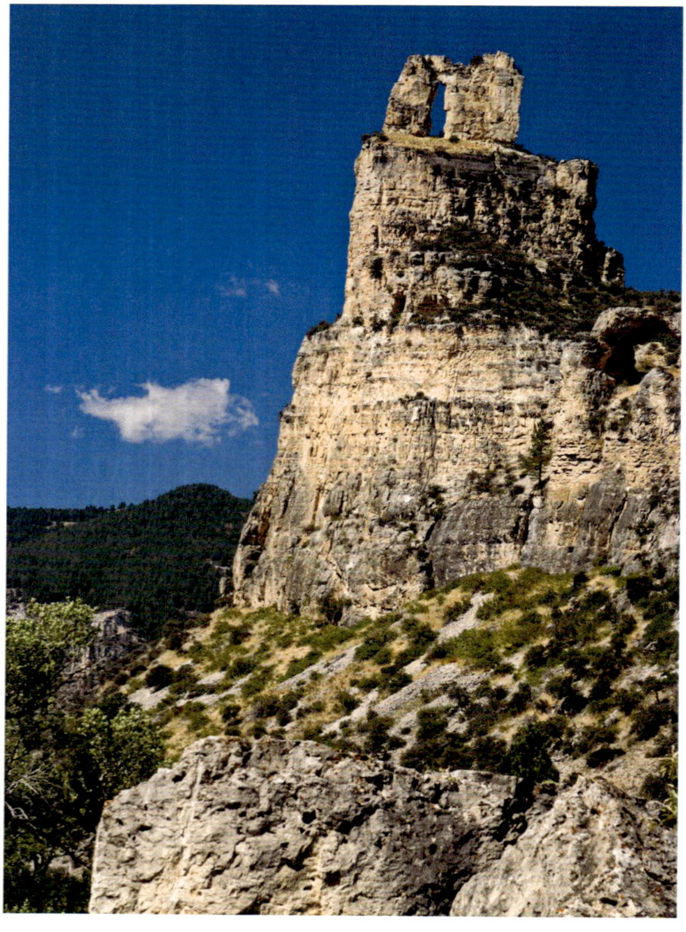

Figure 3-7. *Eye of the Needle cannot be seen on maps or from any road; you have to be told it's there.*

In Alaska, the same technique – asking someone what's worth seeing, a place that's special to them – worked to perfection. Time after time. Only once did someone say "that's for me to know and you to find out."

Inquisitiveness is absolutely key to finding hidden locations. Inquisitiveness (call it ravenous curiosity if you prefer) generates the urge to undertake research, the willingness to talk to people, and the drive to go past "road closed" signs and see what lies ahead. It's simple, really – you don't find hidden locations on well-traveled paths. You have to explore where others rarely go.

In Canada's Northwest Territories, shortly after crossing the 45th parallel, I yielded to temptation, swung off onto an unmarked road, churned through the mud and ignored the alder branches scraping my paintwork. I saw what looked like a large puddle about 100 feet ahead – and, minding its own business, an enormous beaver.

The resulting shot was murky, but the creature was enormous, nothing like its cousins in the States. Definitely a hidden location, and definitely a rewarding one.

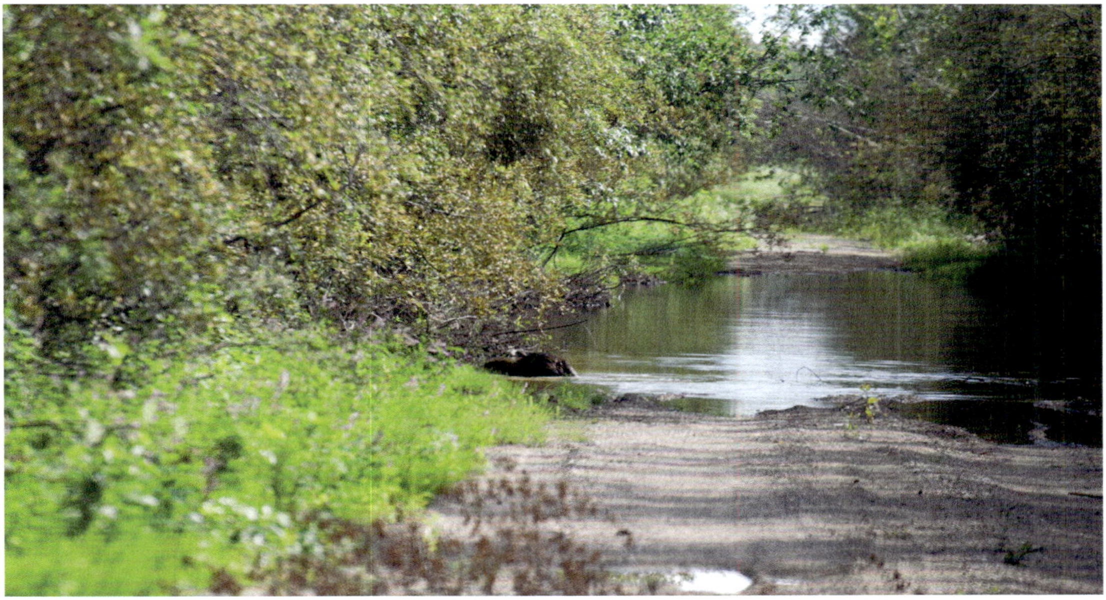

Figure 3-8. *Beaver, just NE of 45th parallel, Northwest Territories, CA. Not the best shot of a beaver, but the neat thing was seeing it in the first place.*

Inquisitiveness was what got the elephant its trunk (he suffered from "'satiable curiosity," and the crocodile pulled his nose until…). Inquisitiveness can get you into real trouble. Being inquisitive is a good thing, but being rash is quite another. But you won't find hidden locations if you're afraid of a modest amount of risk. (See "Managing Risk" in Chapter Two). I didn't attempt to ford the "puddle," just backed out instead. Risk managed!

Effort and perseverance really go together. If you're going find a hidden location, you will be going to some little-frequented place. This will take effort, because if it were easy to get to, everybody would go there. If it were safe and easy, it would be thronged every weekend. So effort is guaranteed. You will have to drive terrible roads, climb slippery rocks, whack through blackberry thickets, dodge copperheads and rattlers, and get a bell to wear on your hat to warn the bears and moose that another stupid photographer is crashing towards them.

When you've been putting out a lot of effort, dripping with sweat and swatting mosquitos for hours, you need perseverance. Lots of perseverance (polite talk for stubbornness). Here's a for-instance.

I was driving along Alaska's Nabesna Road, 40 miles of gravel, washed-away fords and fabulous views. I hadn't planned on being there, but I'd asked the fatal question – who else should I talk to – and here I was 200 miles later in mid-afternoon,

I'd stopped to photograph the tundra and encountered a fellow on an ATV with a rifle strapped to his back. I inquired about the man I was going to see, and was given directions. Then he said the fatal words: "if you drive about 10 miles past the Sportsman's Paradise, the road peters out, but there's a mine worth seeing, and a terrific panoramic view if you get above the trees."

I did have the sense to tell the fellow at the Sportsman's Paradise what I was doing, and promise I'd be back later to talk with him.

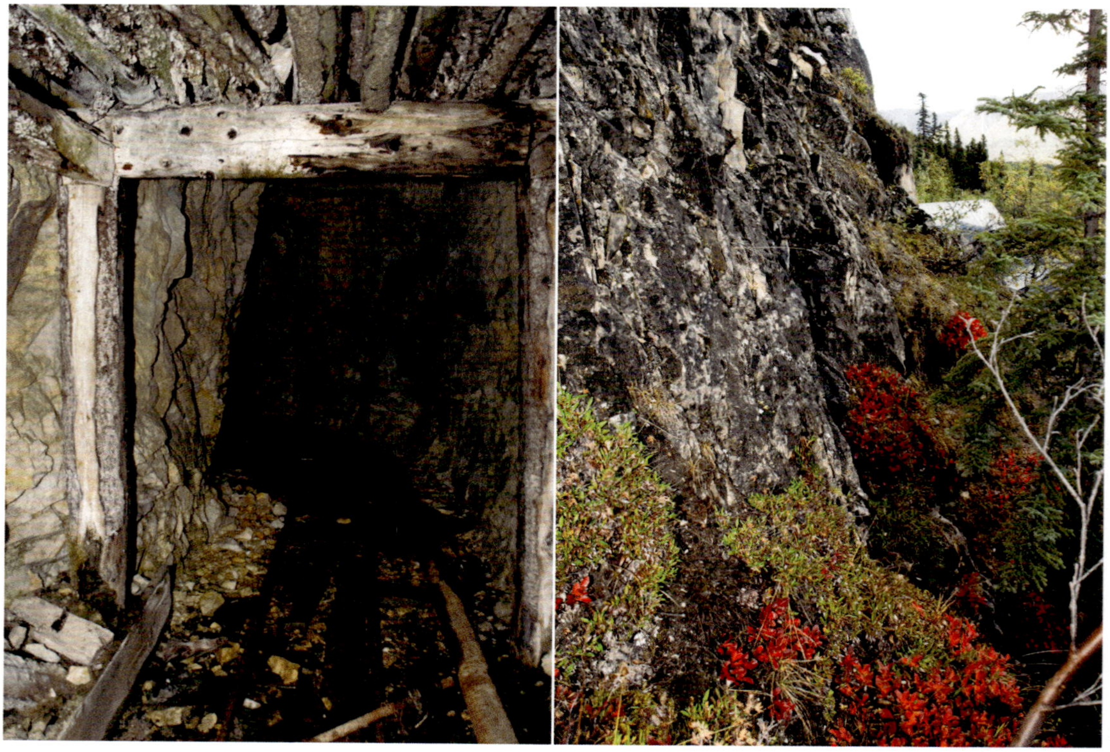

***Figures 3-9 and 3-10**. 3-9 shows the entrance to the now-abandoned "Rambler Mine". Figure 3-10 shows the rugged climb above the mine to reach a promised "wonderful panoramic view"*

It's probably a good thing I was under-prepared and didn't have a contour map. The 15 miles to the end of the road were exciting and incredibly beautiful, with rain squalls and sun shafts chasing each other across the tundra. I pushed on past the official end of the road until it looked too dangerous to continue. Then I saw a trail, took it, and started hiking up.

Being under-prepared, I had no idea what I was getting into. The path rose 800 feet in about a kilometer, which is about a 25% gradient. Temperature around 50, humidity 90%. Ginger (my dog) and I kept climbing. At age 65, I was huffing and puffing, but kept going. Found the mine (Photo above left). But where was the famous panorama? As it turned out, I had to climb above the mine to get above the scrub birch and willow to see. The picture above right shows where I had to climb to. The picture below (Figure 3-11) shows what I found!

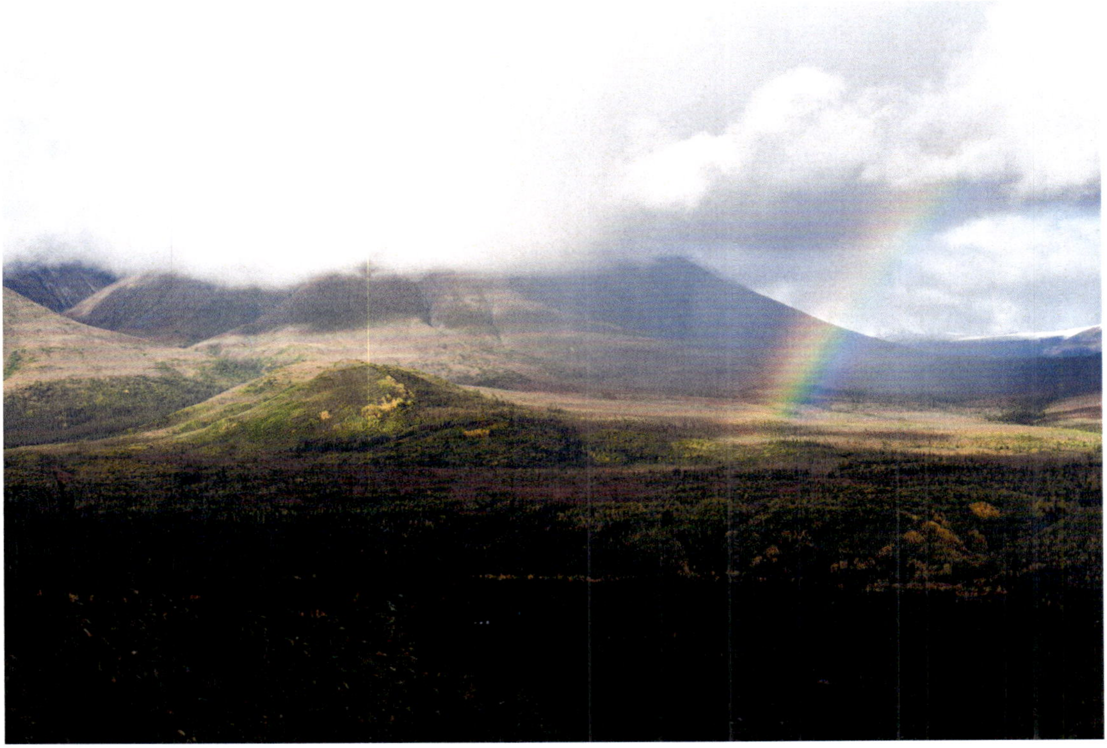

Figure 3-11. *View from ledge above Rambler Mine*

This is, without a doubt, a hidden location. I found it by talking to people, inquisitiveness, effort, and persistence.

A final footnote: I learned later that I could have followed the trail to the mine another 1/2 mile and it would have curved around the shoulder of the mountain eliminating the nasty scramble up the rock face (under prepared, right?). And when that night I caught up with the chap on the ATV, he asked if I'd made it to the mine. "Yes," I gasped. "Why didn't you tell me it's such a hike?" "Oh," he said, "I just find a way around the no-ATV posts and drive right up!" Arrrgh!

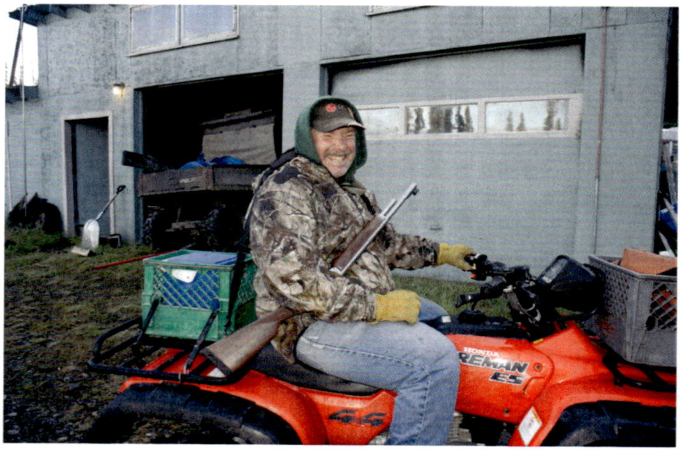

Figure 3-12. *The fellow who told me about the Rambler Mine ... and his reaction when I told him about the hike!*

The final thing you need to find hidden locations is luck. You can't make it, you can't earn it, you don't necessarily deserve it, and you can't count on it – but you definitely need it. And as Gerhard Neumann, the famous aircraft engine designer, used to say, "the harder I work, the luckier I get."

ERIC'S RULES FOR TRAVELING

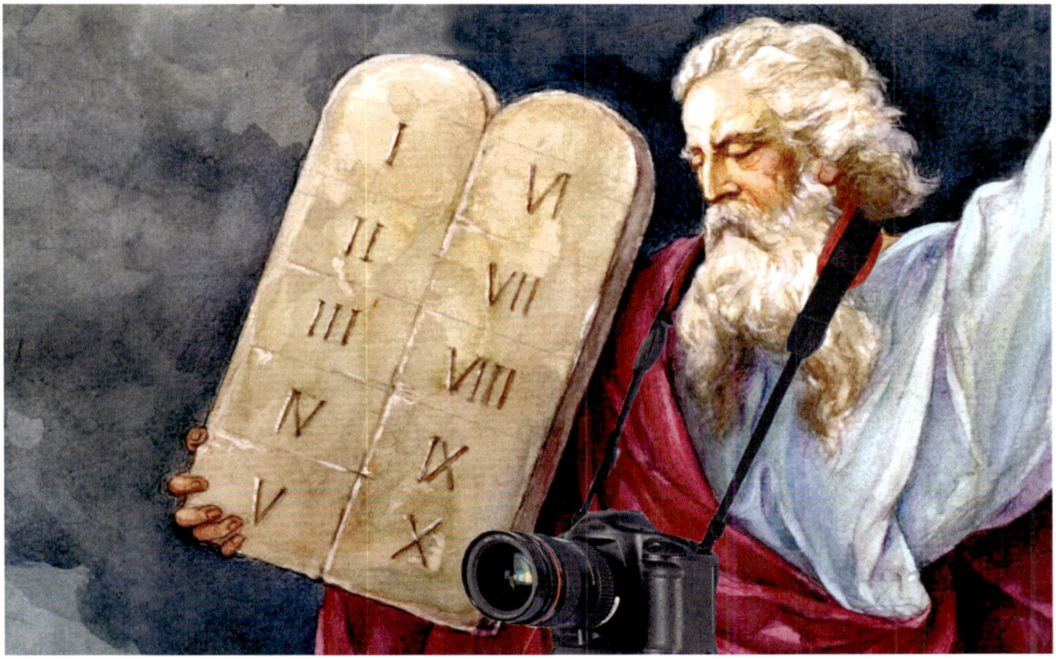

Figure 3-13. *The rules*

I started traveling at the age of three and haven't stopped yet. I've been (and worked) on three continents, all 50 states, and heaven knows how many countries. My first extended trip on my own was a summer spent on in France at the age of 17. I had a great time and learned all sorts of things not on the approved list of educational experiences ... the most important being I could survive on my own when things got tough. (The next most important was to avoid white wine and stick to red, but that's another story).

In the past 5 years I've made extended trips through New England, the American west, northern Canada, and Alaska, plus some time in Hawaii.

All of this has led me to some rules for successful traveling. They apply broadly, but ESPECIALLY to photographers.

Rule 1. *Know your objectives.* What are you traveling for? The answer can be anything you want – see the Alps, swim in an ocean, photograph a glacier, meet the locals — but it helps to have some objective to shape your trip and keep you going.

Rule 2. *Research your destination.* The internet has made this a snap, but it helps in setting your objectives and giving you some clue as to what you're letting yourself in for.

Rule 3. *Don't over-plan.* Nothing kills a trip like trying to do too much. "If this is Tuesday, it must be Belgium" really sucks because you won't be able to process your experiences. If you commit to too many stops, too many locations, you'll be frazzled and disappointed when it rains, or the road is closed, or the bus is late,

When I travel, I like to make a list of things I want to see — and give myself permission to skip any or all of them if something really interesting takes priority. Dwight Eisenhower said, "Plans are useless, but planning's essential." He was right, because circumstances force plans to change – and the more detailed the plan, the more likely it is to go wrong. Avoid over-planning to build flexibility into your trip.

Rule 4. *Use tours judiciously.* Tours can help you or louse you up. If you have little time and no knowledge of the place you're going to, tours are valuable.

Tours can familiarize you with your destination. They can also take you places you simply couldn't reach (or understand) on your own. In Alaska last summer I took only two tours: one through Denali (you can't really do it any other way) and one from Stewart for a wildlife and glacier cruise. Worth absolutely every penny!

But for travelers who want to experience a place, not just see it, tours are only useful for quick overview. Then the next day wander around or revisit some of the places you craned your neck to see from the bus. The other main drawback to tours is the limited time you spend at a stop. Photographers frequently need far more time than the tour guide wants to spend.

Hiring a local guide for the day may be affordable and is a great way to get access and local knowledge without jamming yourself onto a tour's schedule. Worth considering – even if you have to team up with a couple of other travelers to keep the cost down.

Rule 5. *Make sure you can meet local people.* They are your best resource for finding great places to eat, great places to photograph, and great places to avoid. But how do you do this?

1. Sleep in pensions, campsites, hostels, inns, farmhouses, or B&Bs rather than large tourist hotels. The place you wind up in may be splendid or perhaps less than that, but you're trading off luxury for experience and local knowledge.

2. Go to the bar and strike up a conversation. Buying a drink or two makes it easy!

3. Ask at the local tourist bureau or welcome center to suggest good spots to meet locals. I did this in Canada and wound up a day later on a free ATV trip back into some mountains, along with a really interesting conversation with some First Nation folks, and the most incredible dog I've ever met – but that's another story.

4. Ask these two questions. They work every time.

a. "You're from around here. What should I photograph (or see) that isn't on every postcard – maybe some place that's special for you?"

b. "Who else should I meet while I'm here?"

These two questions are the absolutely best ways I know to make sure you have a wonderful trip. (See The Marvelous Act of Asking Your Way elsewhere in this book).

Rule 6. *Don't over pack.* This is a tough one, particularly for photographers, as we NEVER travel really light – the gear weighs us down. But make the effort – every ounce you pack is an ounce you have to schlep, and they all add up! My rule of thumb is pack so that I can carry it all at one time. This means minimizing separate pieces by doing things like making sure your tripod fits into your rolling duffle, and that your lenses are multi-purpose rather than primes.

The truth is you WILL over pack. But truly, it's worth making the effort to minimize. Really minimize. Just don't forget the toilet paper if you're going into the wilderness!

Rule 7. *Figure out how to keep your images safe.* I usually travel with a laptop and a portable hard drive. Every night, religiously, I empty my memory cards onto the laptop and duplicate them onto the hard drive. Then I pack the laptop and the hard drive in separate locations. I usually travel on my own so I can find time to edit out trash shots while I'm at it, thus saving space.

Rule 8. *Take good notes.* I use a digital tape recorder for this ... I can record locations, impressions, etc. very easily for later use at home. Keeping a geotag log is a great idea. Your smart phone can even do this, and so can many cameras. External geotag units are available as well. But there's nothing like being able to pinpoint when and where you took that superb shot of a golden eagle swooping up a gopher in the setting sun!

Rule 9. *Consider your power needs and get whatever you need before you leave home.* You have to make sure you can recharge or replace batteries, get online if you have to, have enough memory cards, etc. Verify what power you need (110v or 220v) for instance, and match it (or get a converter rated not to fry your gear) to the power grid where you'll be traveling. We live and die by electronics these days, so make CERTAIN you are covered in this regard.

Rule 10. *And one of the best.* Never be afraid to ask for help if you need it. Like asking your way, asking for help appeals to peoples' best sides. The converse is also true – if someone needs help, be the first to step up and offer it You may make a friend for life, and you will certainly enrich your karma by doing unto others.

LEMONADE DAYS AND MORE

I've taken many, many trips in a long and vigorous career. A few of these stand out, and I've spent a little time thinking about what made them work. I've written about this topic in "Eric's Rules for Traveling," but there's more to be said.

The best trips have had a few specific goals, and plenty of flexibility about how to reach them or to change them if need be. In 1962 I spent the summer in France at the age of 16. I bought a used moped, got a Michelin Red Guide, and formed the general plan of touring counter-clockwise and getting as far as I could. I definitely wanted to see Chartres, Nice, Monaco, and the French Alps.

This plan actually worked! (Except for perpetual moped breakdowns and getting thrown in the slammer in Marseille, for improper documentation for said moped. But that's another story). It gave me a few definite targets and plenty of opportunity to go see what was interesting. And at that age, everything was interesting; I knew virtually nothing except how to open a wine bottle and how to speak pretty good French.

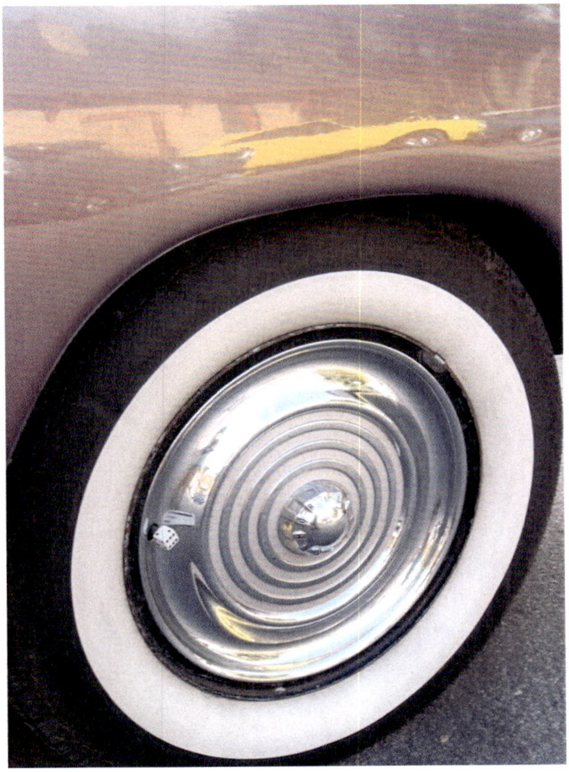

Figure 3-14. Famous old car show in Deadwood Gulch, ND. Note the lucky dice valve cap. I had my share of the luck.

That same basic model worked for me out West in 1007 and even better in Alaska in 2010. In 2007 I wanted to see the Badlands, and that was about all I had in mind. I added destinations / and sights to see based on what I learned from local people. This worked great! For instance, from the lady who ran the Rocket Motel in Custer, SD, I learned about an upcoming old car event in Deadwood --- got great shots and won $400 the casino (enough to pay my dog's vet bill, but that's another story). Note the air valve on this old gem!

On this same trip, on the way to Deadwood, it poured rain. I'd had to disappear into the woods (bad water or something) in a downpour. Then, despite myself, I found I had to go past Mt. Rushmore, one of the most over-photographed spots in the US, and one I'd sworn not to visit. But here I was ... and because of the rain I got Figure 3-16 using my open tailgate as an umbrella.

This was a lemonade day! That's what happens when God tosses you lemons, and you make lemonade out of them. Washington was weeping, and how many tourists get to capture THAT?

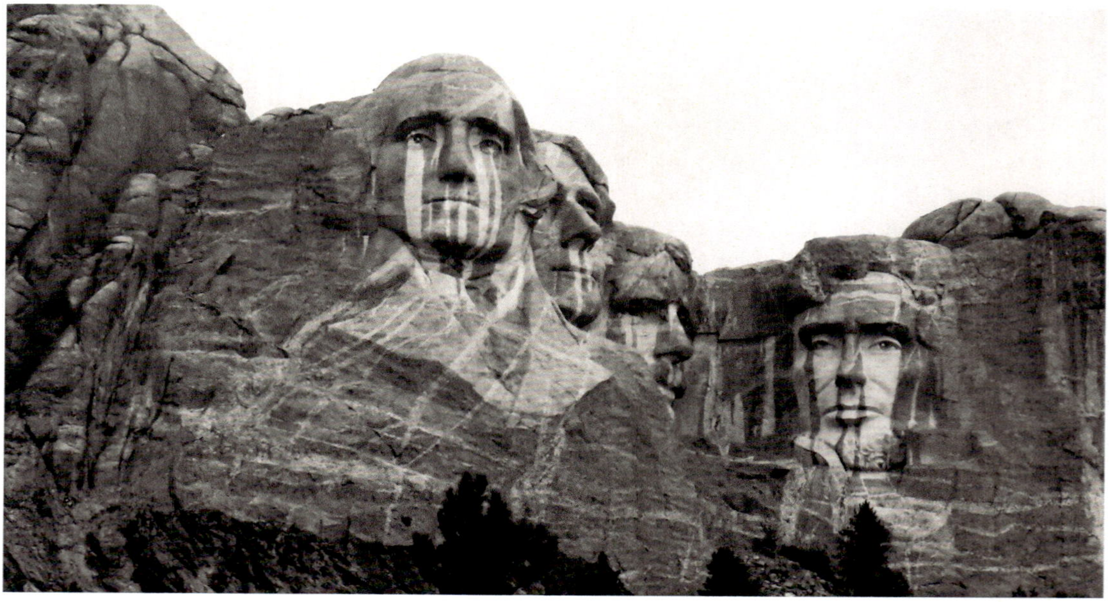

Figure 3-15. *Washington weeping on a rainy day at Mt. Rushmore.*

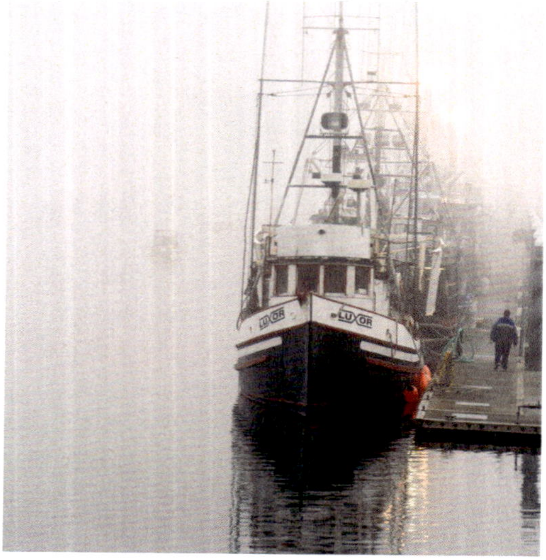

Figure 3-16. *A lemonade day in foggy Washington State*

I have learned how to make lemonade. On my 2009 photo safari to Washington State, I had two specific objectives. Weather and circumstances interfered with achieving either one. For instance, we spent a morning when we had planned to be on the water waiting out fog and repairing the boat motor. But I used the opportunity to shoot bald eagle pix and to take a couple of really nice boat-in-the-fog images.

This was a lemonade day -- got handed lemons, made lemonade, and it tasted pretty good!

Lemonade days can have some surprising effects. They can force you to slow down, to be patient, to find a spot to just be quiet and think while you're waiting for the fog to lift, the sun to come out, or the buffalo to come through the valley. Lemonade days and also invite you to shoot stuff you don't normally go after: a wet spider's web, a crowded café, an abandoned factory … pretty much whatever you weren't looking for, but because you can't do what you planned, you can suddenly see a new opportunity. Lemonade comes in many different flavors!

THE MAGIC OF GRAVEL ROADS

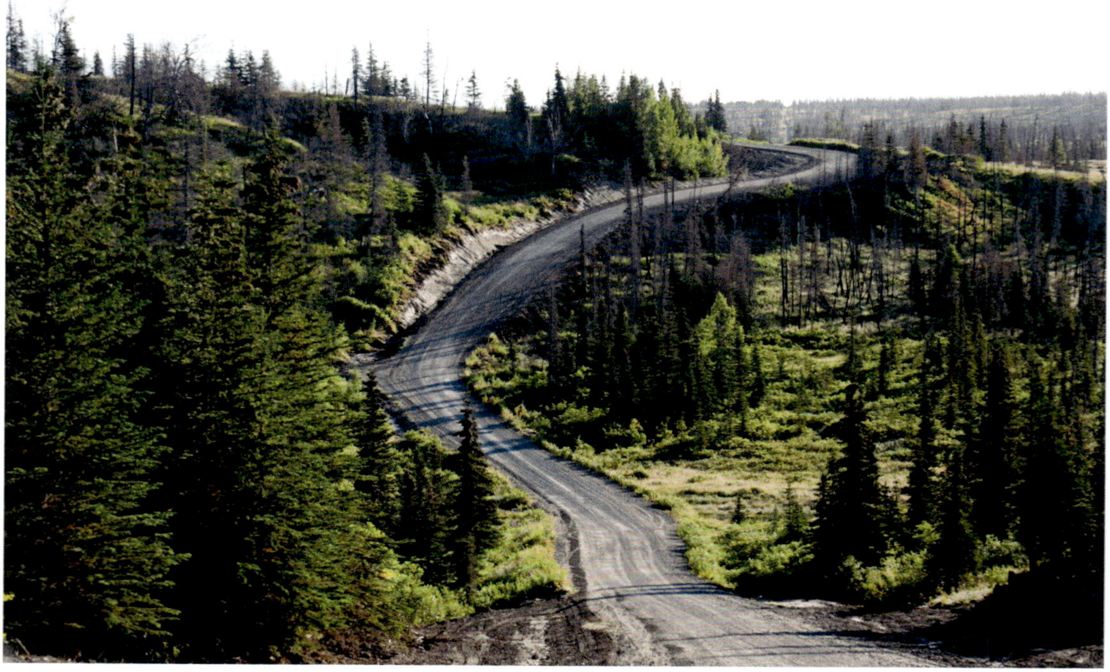

Figure 3-17. *Oil Well Road, Ninilchik, AK*

My love affair with gravel roads begins in the gritty New England of my youth. In those days northwestern Connecticut was a land of dairy farms, rock-strewn cornfields, and hockey on the village pond. The Litchfield Hills were long wooded ridges with a stream in every valley, laced together by an intricate network of dirt roads. The nearby Berskhires were steeper and more rugged. A teenager seeking a little romance with his girlfriend quickly drew a detailed mental map of every dirt road within 50 miles of home … and was boastful of his knowledge.

The gravel roads of New England were a hoot to drive, except in mud season, when they became nerve-sapping mires. I remember a sign, back before all of Vermont 100 was paved, that read "choose yer ruts careful, you'll be in 'em a while." And oh, 'twas true, and thank God for newish snow tires. But mainly, driving these roads with their rocks, washboards, and potholes was just great fun. You never knew what you'd see — an abandoned village, a mountain lion (yes, in 1955 or 56 I truly did see one springing from a boulder and leaping across the road in front of us).

My father, a well-known humorist and novelist, loved driving – it was where he did his thinking about the next day's stint at the typewriter. If we were quiet, my brother and I could go along for the ride. The Colonel (as he was called by many) frequently had no idea of where he was in the physical world, being fully engaged in the world of his imagination. But as kids, we never worried. Dirt roads invariably did one of two things: they emerged onto a larger road, or they petered out in a farm yard or trail head, in which case you just turned around and waited til Option One ("coming out somewhere") occurred. Satellites had barely been invented, GPS lay five decades down the road, and we had a perfectly wonderful time "getting lost."

These expeditions bred in me a genuine love of driving for the fun of it, for the adventure of finding something new, and for the occasional challenge of extricating yourself from any number of unforeseen situations. (I vividly remember backing up 5 miles of logging road in New Hampshire, and doing the same for about 2 miles of mountain shoulder in Jamaica 10 years later). A few years ago I had occasion to drive through much of the Western States, and last year spent two months on the gravel in northern Canada and Alaska. This gave me plenty of time to reflect on the magic of gravel roads. Here are my conclusions:

Blacktop roads and interstates urge you to get from A to B as quickly as you can. They're all about the result, a quick arrival, and minimum time on the road. On an interstate, you see virtually nothing but the semi in front of you – a boring view to say the least.

Gravel roads issue an engraved invitation to satisfy your curiosity. Where does this road go? What's around the bend? What lake is that I'm glimpsing through the trees? Gravel roads trigger the imagination. They're seductive, adventuresome, teasing.

Where concrete induces boredom, gravel incites a broad range of feelings. Gravel roads may intrigue, beguile, terrify, excite, and even flat-out astound you with some unexpected beauty.

Most of the time, especially in hilly or mountain areas, gravel roads follow the contours of the land, and because you're going slowly, you have the opportunity to look around. New views are around every curve, and there are LOTS of curves. Where asphalt says, "hurry," gravel says "slow down and enjoy the show."

TRAVELING AND SHOOTING WITH YOUR DOG

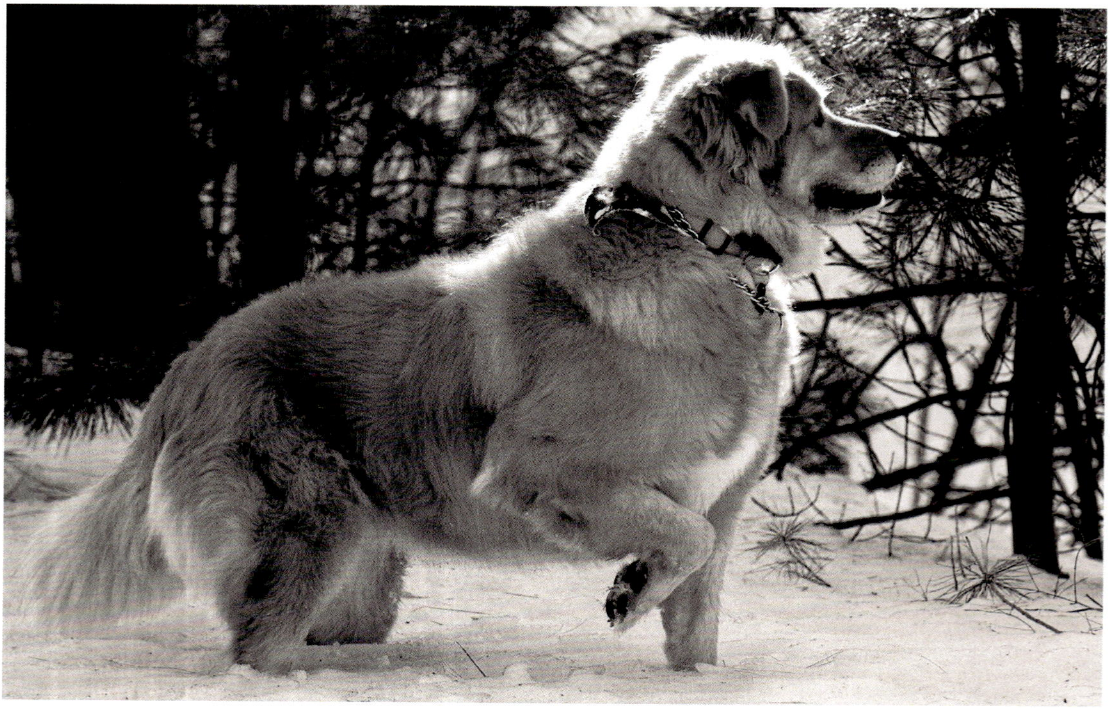

Figure 3-18. *Ginger in the New Hampshire snow.*

Traveling with your pooch is an adventure. Whether it's a great adventure or not depends on lots of things – where you're going, your dog's personality and needs, and on what you're trying to do on the trip.

My trips are photo expeditions to isolated places. I load the car with camping gear, food, two or three cameras, tripod, clothes, miscellaneous supplies – and my 65 pound Golden-Chow cross, Ginger. We alternate camping with cheap motels.

So here's the Letterman list of the top 10 reasons I travel with my dog:

10. When your wife, your photographer buddies, and even your daughter not only won't go but also argue against the trip, all you have to say to your dog is "Let's get in the car." ZOOM. Tail wags, eyes light up, and you're NEVER going to get an argument.

9. You don't have to pack any extra cosmetics, sports outfits, or coats. You DO have to pack dog food, a dish, and a leash. Dog food does take up space, but the good news is that you can always eat it if things get tough enough.

8. Depending on your dog, you have a built-in conversation-starter. Ginger has an aura that I absolutely lack – she can raise the warmth of an atmosphere faster than Santa Claus at an orphans' feast.

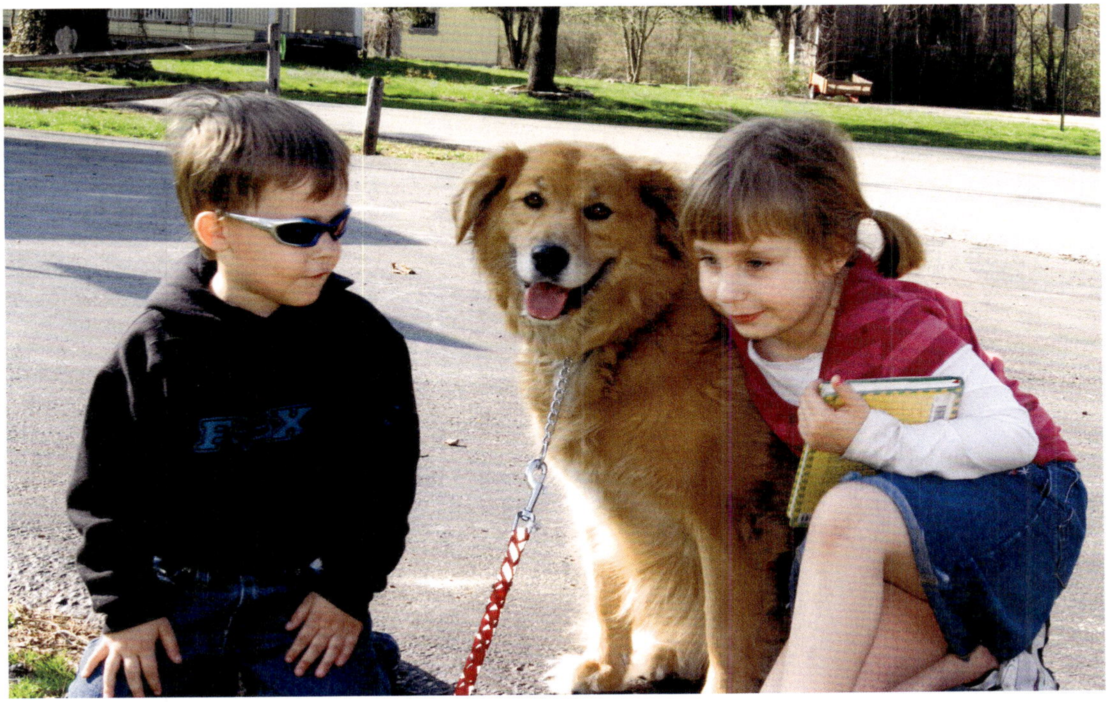

Figure 3-19. *Your pet can help you make friends anywhere.*

Ginger really does evoke the coo-and-cuddle response among both men and women (more women, though). If your dog is a Chinese Crested Hairless crossed with a mastiff, you might not get the same reaction. But Ginger makes an ideal door-opener – sometimes literally.

Ginger's ability to open doors was proven one night in Red Lodge, Montana, when we arrived late, tired, and having had a hard time finding lodging. Eventually we did find a room. We found a parking space, and Ginger surged into what she figured was our room. OOPS! There was a Harley parked in front of her choice, an open screen door, and in she went.

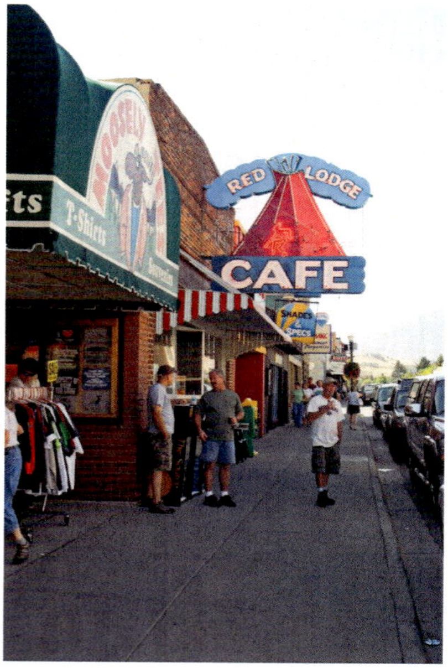

Figure 3-20. *Red Lodge, Montana.*

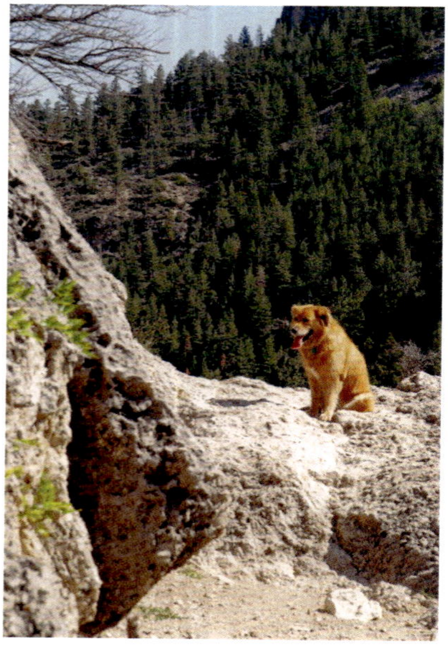

Figure 3-21. *Your dog can add scale and depth to a shot.*

I followed quickly, apologizing all the way. What I saw was a blonde in the bed and a middle-aged guy with a ponytail and a 3-day beard watching the TV. Uh-oh!

That's when the blonde pulled up the sheet, looked at Ginger, and cried out "Ooooh, what a sweetie-pie. What a good girl!" Well, there are door-openers and door-openers!

7. While conversation with your dog is apt to be limited, she won't argue politics, won't accuse you of hating her mother, and will always listen to you. In fact, if you're traveling without a human companion, your dog can just plain be good company. Good reason to take it along.

6. Your dog can help you get out of trouble. Ginger is very good at following trails that aren't marked well or are over rocky terrain. Mostly, I don't need help, but there was a day in Maine last summer where her nose really did make getting back faster and safer. Of course, she could have been following a squirrel, but you just have to take your chances.

5. Your dog makes a great car alarm and burglar deterrent. The bad guy doesn't know how loveable your mutt is. He just sees all those teeth and thinks twice.

4. Your dog probably hates kennels, and good ones cost more than motel rooms these days. Save the money and take your dog with you.

3. My dog makes a terrific prop, scale-gauge, and perspective-generator. Put her in the frame and there's suddenly a sense of place, an energy, that wasn't there without her.

2. Dogs like other dogs (mostly). This leads to some fun socializing. Your dog will make the introduction for you! I'd never have met this extremely attractive young French woman in Framingham, Mass without Ginger to enable the conversation. Ooh-la-la.

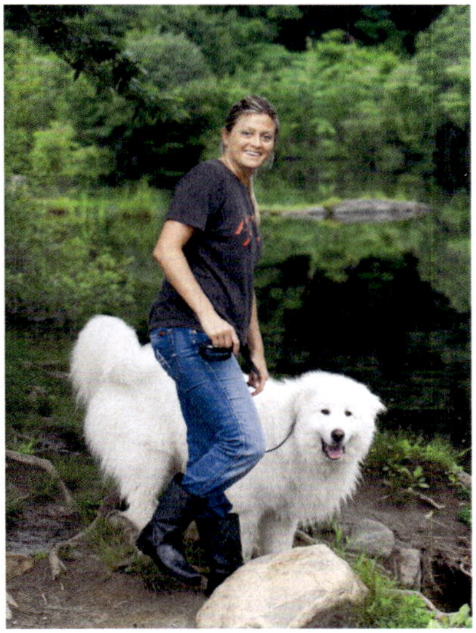

Figure 3-22. *Your dog can bring you lucky encounters*

1. Maybe the best reason of all – your dog will love being with you, sharing your adventure, and maybe even carrying a spare water bottle or your sandwiches in a back pack of his own. There's just nothing like it.

Now the negatives:

10. Dogs shed. Boy, do they shed! As a photographer this means you will find dog hair in your trunk, inside your clothes, on your lenses, even inside your camera. It's a pain … but you have to clean your gear often, because otherwise you're inviting sticky autofocus, doggy sensors, and lost photos. Thanks, Rover!

9. You have to allow for your dog in your planning and arrangements. If you're staying in motels, you need to find ones which permit pets – or learn how to sneak yours in and out. And in. And out! Non-chain motels tend to be a lot more pet-friendly than the biggies, as well as cheaper.

8. Your dog can tie you down and limit your freedom of movement (just as you may have to limit his). In National Parks, dogs must be leashed and kept on trails. In Denali, they must be left in campsites, and the only dog-sitter in the area has just moved to Talkeetna and gotten her pilot's license. Yellowstone is even worse; it's a canine gulag. If you're a photographer like me, those rules can make a difference. The restriction is a nuisance, but it actually makes sense, protecting endangered species, such as the delicious Western Marmot, or the toothsome Kansas Prairie Bunny. It may also keep your dog from being lunch for larger, meaner critters with less genteel manners than your dog has. Ever seen an eagle lofting away a Papillion? It happens.

7. In France, dogs are welcome in taxis and in restaurants. Not in this country (with rare miraculous exceptions). That means when you go inside to chow down, Fang is left somewhere else. But where? Broiling in the sun? Howling his poor deserted lungs out while you're gone? So again, we're back to extra planning.

6. Your dog can genuinely get in trouble. If your dog gets lost a thousand miles from home and a million miles from nowhere, you're got a real problem (hunters can now equip their dogs with GPS trackers, but that's a pretty drastic approach). If she swallows a forbidden chicken bone, as Ginger did in Custer, South Dakota, you have to find a vet and pay the freight – or not. I now carry pet medical insurance to cover this kind of thing. Not cheap, but the vet bill in Hot Springs SD (where I finally found a vet) turned out to be $400 WITHOUT surgery.

5. Dogs have to be walked. Where we live out in the country, this is a matter of opening the door and saying "Out, Dog." But you can't do that on the road. So if your dog is nervous, and announces at 3:00 AM that she HAS to go out, shuffle into your clothes, put on the leash, and open the door into the soaking rain. And keep hiking until Spikette finally relaxes enough to let you go back inside. NOT the best way to get a good night's sleep. Oh, and don't forget the plastic poop gloves.

4. Your dog can distract you from the business at hand. It's really hard to concentrate on the photos you've driven all this way to take if you have to keep an eye out for your wandering pal. When Ginger had only been with me for a few months (she's a pound rescue), I was working outdoors with a model in a steeply gullied area. The model wasn't wearing much of anything, and the lighting conditions were difficult. So quite obviously I was NOT watching Ginger. When we finished the shoot, we discovered that Ginger had vanished! My model and her boyfriend and I hiked the gullies for a good two hours. God bless 'em, they wouldn't drive back to their digs in Cleveland til they knew Ginger was safe! (See, that's the ole doggy magic thing).

   She had found her way into a housing subdivision and been instantly adopted by a family with four other dogs. (Doggie magic one more time). They eventually called me, and we were reunited. But since that time I've always had to keep from full submersion in the photo at hand. Out come the leash and the complaints that go with it! And in her favor, Ginger has learned not to wander while she's with me. But still, if you're dog watching, your mind can't be wholly on the job.

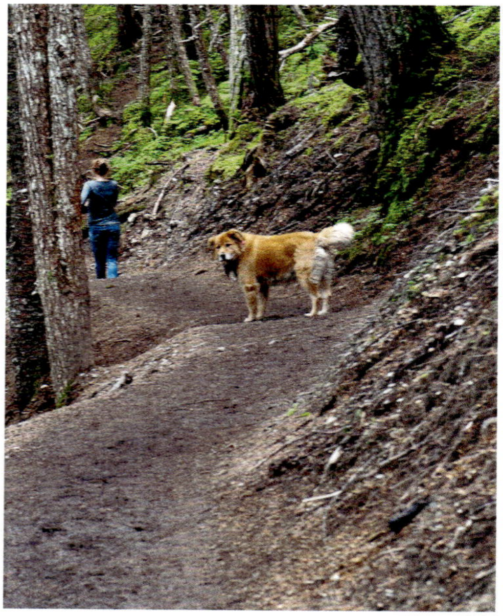

Figure 3-23. *Come on, boss, what are you waiting for?*

3. Busybodies can make your life hell. If you stop at a rest area and trot inside for a quick stop, you will come back out to find a cop staring into your car, or someone from the local SPCA is phoning your plate in for animal abuse. Doesn't matter if windows are open, the dog's just been watered, and the outside temp is only 70, SOMEBODY out there is convinced you're the Simon Legree of dog owners. Just expect it.

2. Your dog's sense of smell differs from yours. Stuff they don't mind, or even seem to like, can cause major problems when there's no handy dog wash and you're in a tight space like a car.

   I mentioned that Ginger had eaten a chicken bone. The vet got it out by a combination of depth charging and power-flushing her, to put it kindly. It was that or major abdominal surgery. You can imagine what that smelled like. Two days later, when I picked her up, they'd cured her but hadn't washed her! I drove around Hot Springs SD with the windows wide open til a competing vet took pity on me (the town's only dog groomer was gone someplace) and agreed to hose her off.

Skunks are another hazard. Likewise road-kill. Why do dogs like to roll in 3-day old deer guts? I remember fondly the night Ginger climbed into the tent with the foreleg of a deer – and crunched it all night long.

1. It's 10:00 clock at night. It's cold. You've walked the dog, who now smells like a wet blanket coated in fermented anchovies. You're no longer hungry. It's been raining all day, and you just don't want to pitch that tent. You scramble into your sleeping bag in the back of the SUV. You are just falling asleep, when OMG! Your dog has just cut loose with a Black Stealth Fart, somewhere around 9.0 on the Fart Richter Scale. Don't you love traveling with your dog?

CHAPTER 4: *Understanding Key Technical Concepts*

CHAPTER 4: *Understanding Key Technical Concepts*

BLACK AND RED AND OUT OF FOCUS (WHY YOUR AUTOFOCUS LETS YOU DOWN)

This entry is an expanded version of a response to a query from a photographer made in the discussion forum of the Ohio Valley Camera Club. The shooter wanted to know why she couldn't get good results using autofocus shooting a red object on a black background.

Most of the answers had to do with distance from the object, but there's more to it than that. Part of the problem is the way autofocus works (passive autofocus, at least, which is the kind you get on good digital cameras). And part of it is the mysterious kinship between red and black.

Autofocus works by detecting an object in the frame and measuring the degree of contrast between it and something next to it. It now makes a small change and sees if the contrast has become greater. When it's at its greatest, the autofocus says "voila!" and stops adjusting. If it can't find sufficient contrast, it keeps hunting or gives up in disgust.

The more contrast, the easier it is for the autofocus to do its job. In scenes of low light, there's very little contrast. In scenes with a monotone subject, such as snow on the ground or clear blue sky, it's also hard for autofocus to find enough contrast to grab a target. Same reason; not enough contrast for the autofocus sensor to grab hold of.

You can read a basic explanation of the autofocus process in How Stuff Works and a better one here (Cambridge Colour.com)

Now remember that we were talking about a red object on a black background, such as this one in Figure 4-1.

This image consists of pure red, exactly half way between full light and full dark (128 on a scale of 0-256) on a black background. That's represented by the thin red line on the histogram shown above the red square. It's very short, but you can just see it. On the left is a longer line all the way at the left edge. That represents how much of the image is dead black. No problem for autofocus, right? Well, it IS a problem, and here's why. Autofocus sensors do not see color. They only see black and white. And when you desaturate this image, you get Figure 4-2.

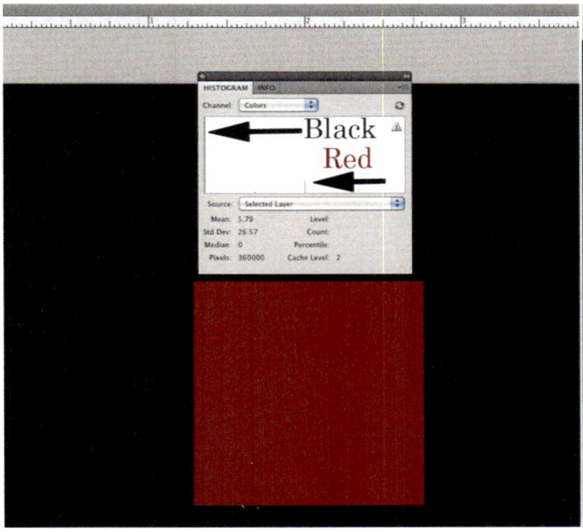

Figure 4-1. *The length of the red line shows the amount of red compared to black in the whole picture. The position of the line shows where red falls between light and dark.*

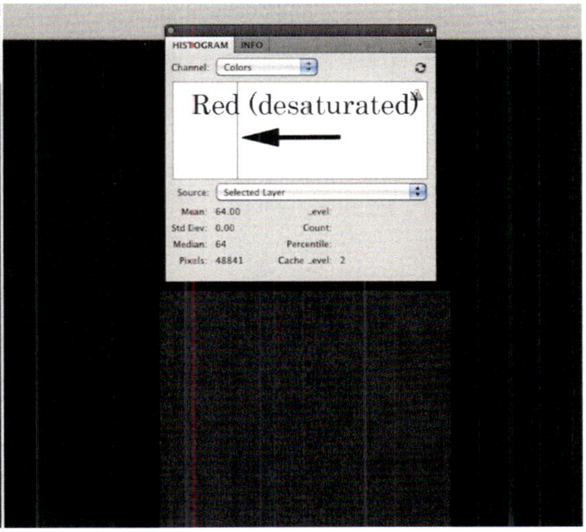

Figure 4-2. *The autofocus sensor reads only black and white info. The luminance value is now much closer to pure black, leaving less contrast for the autofocus to work with.*

When you take the color out of red, leaving only its luminance (light-dark) value, it is nowhere near "middle gray." In fact, as the histogram shows, it's barely half way between pure black and middle gray. So your autofocus sensor actually has much less contrast to work with than our eyes think it should!

Now both these illustrations are for crisply defined edges. What happens if the red object was, say, a soft fuzzy pillow? Here's the same red-on-black saturated image with the edges blurred:

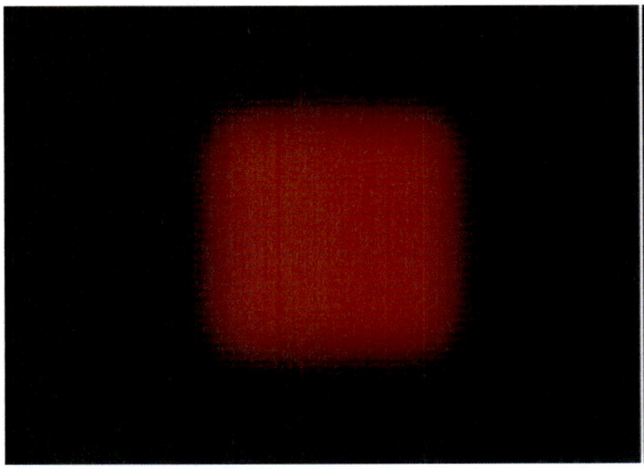

Figure 4-3. *If there are no sharply defined lines in the picture, the autofocus sensor is doomed to failure; it simply hasn't enough contrast to work with.*

Figure 4-4. *Shows luminance only; not much to work with. Your autofocus sensor is BW.*

Now there's even less contrast to work with. Your autofocus would have a tough time with this image. You can tell the autofocus is in trouble when you hear it whirring and whirring and the camera tries to find an area of sharp enough contrast to focus upon.

Sometimes autofocus can't handle things because they are too close together, or because the high-contrast region picked up by the sensor blocks out the subject you DO want to have in focus. No autofocus system could handle this image of a girl behind a screen.

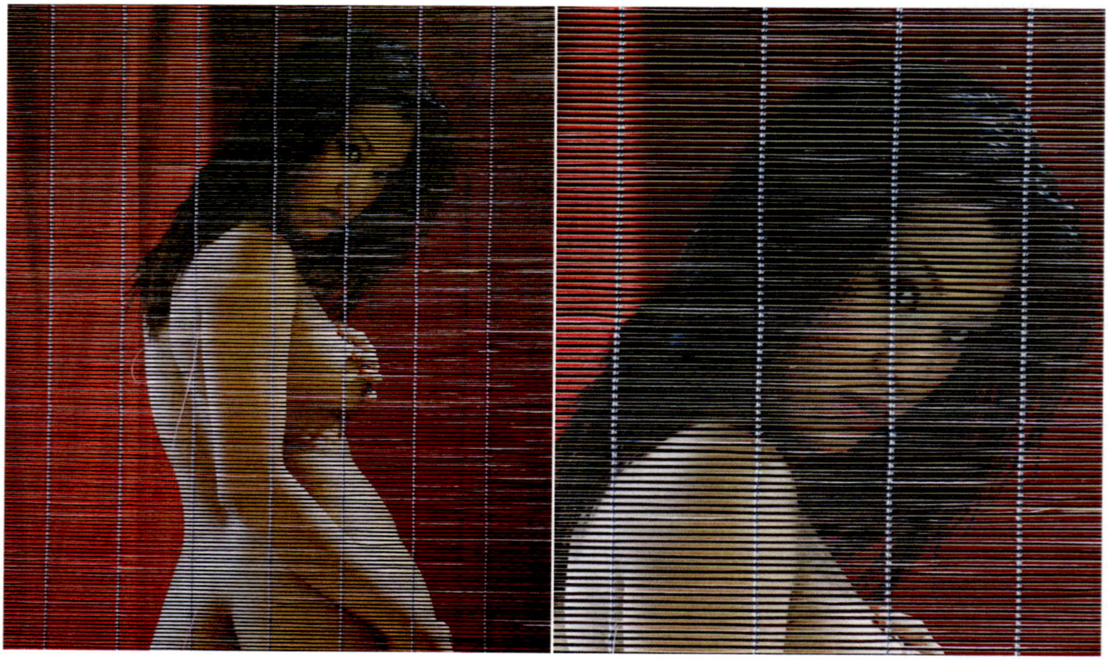

Figures 4-5 (full frame) and 4-6 (crop). *The bamboo screen will "distract" the autofocus sensor, because it has more "contrasty" areas to work with.*

At 100% crop (of a downsized jpg, so it looks lousy) you can see that the girl is actually in focus while the bamboo screen is not. This HAD to be done manually, because autofocus would grab the screen and not the model. I had to raise the screen, manually focus on the model, lower the screen, and take the shot. I had to do this for every exposure because she'd move between exposures, and I was using an open lens aperture with narrow depth of field. We were both beat when this session was over!

So, to sum up, autofocus looks for sharp edges (high contrast between one pixel and the one next to it). If it can't find them, it goes hunting or grabs the wrong part of the subject. That's why autofocus may grab a leafy branch instead of the face in front of it. Checkerboards and such are "autofocus magnets," while soft fuzzy things without much light-dark contrast are hard for it.

BASICS OF RESOLUTION

Megapixels are a mug's game – most of the time. If you're producing close-ups of your newborn which will be distributed via cell-phone or web, you don't need a ton of megapixels. For such photos, 3 megapixels would be plenty, even if you wanted to print the occasional 5x7 or 8x10. You don't need a $5,000 rifle to shoot woodchucks (and yes, you can buy such a varmint rifle).

So what's the big deal about 6, 8, 12, 22, or 36 or more megapixels? It depends on what you're trying to do and how the images are intended to be used.

Super resolution only matters when you're super fussy and are planning on making super-large prints, OR when you're planning on cropping to just a small part of your frame and then blowing it up to even normal sizes like 5x7 or 8x10.

Notice, I said the P-word, as in PRINTS. There isn't a monitor made that can portray the full dynamic range and intricate detail of a finely crafted print. That's not an opinion – it's a simple matter of fact. The technology of monitor display spaces color dots too widely and too imprecisely to really capture silk-smooth texture or subtle tonal shifts. You need a high-quality print to do that.

There are a number of components affecting resolution. The most important are glass, cleanliness, pixel count, pixel size, and the camera's internal data-capturing algorithms.

For now, let's leave the camera's picture-taking logic out of it and assume you have a competent camera – today, most digital cameras are better than competent, they're really rather good at finding exposures, autofocus, and the like – all of which affect the final resolution (clarity) of the image.

Glass

The quality of your glass (lenses) really matters. Imagine that you smeared Vaseline all over the front of your lens – no way would a picture shot through than lens ever be sharp. Ugh – yet in fact, smearing Vaseline on a lens was an early way to achieve soft-focus and vignetting ON PURPOSE.

Lenses get very expensive very quickly. That is because one piece of curved glass is never going to do the whole job. So lenses are made of many clusters of glass (elements), each with its own coatings, curvature, placement in the barrel, transitivity (ability to transmit light), and color characteristics. The greater the transitivity (read ability to shoot in low light), the greater the cost of the lens. That's why a 300-millimeter f2.8 lens costs about 4 times as much as a 300-millimeter f5.6 lens.

Dirt

Even a thin film of smoke, chemical deposits (as from cheapo lens cleaning sprays) will interfere with the lens's ability to do its two main jobs: transmit light and focus light without reflection, fringing (chromatic aberration), color distortion, etc. Dust is your enemy – and I know this to my cost from extended shooting in northwestern Canada and Alaska, which are permanently and inescapably dust-ridden.

You can do things to reduce dust, but you can't eliminate it permanently. It WILL build up and haunt you. You have to go after it carefully, frequently, and thoroughly. And that includes cleaning your sensor, not just your lenses.

So if you need really high resolution, get the best glass for the job you want to do. And keep it clean, but don't scratch it while cleaning it! And by the way – don't apply a cloth to the mirror in your camera — the mirror is not coated and scratches very easily).

Megapixels (Quantity)

Now, at last, we get to those megapixels. Here's the deal. Let's imagine you are a landscape, architectural, or product photographer. Let's further say you have a 6-megapixel camera. You take a picture of a distant tree with an eagle sitting on top of it. The tree is standing in a rocky field. The tree and eagle take up about 20% of your shot – the rest is field and sky. So only 1/5 of your pixels are available for the important part of your subject! In this example, that means you have at most 1.2 megapixels out of 6 dedicated to tree and eagle. If you wanted to crop the image to show just the branch and the eagle sitting on it, you're down to about 5% of your pixels, or about 300,000 pixels.

You're not going to get any kind of sharpness and detail from that number of pixels.

Let me make this simpler. The more of your pixels are devoted to your subject, the better your print will look. If you have only 20 pixels devoted to your eagle, you're doomed. 20 million pixels – now you can make a 30x40 and it will jump out of the tree at you.

That's why resolution matters much less for close-ups than it does for landscape shots. You don't crop close-ups much, so you don't give away pixels to useless parts of the image. On the other hand, if you want every detail in a landscape to be clear and sharp, you need great glass, correct exposure, and all the pixels you can get. Here are some examples:

This picture of a bear is uncropped. It was shot with a 70-300mm zoom on a 12-megapixel camera through an open car window from a distance of about 12 unsafe feet in a downpour. The subject takes up most of the frame, so there's no problem with resolution. Unless you're planning on very large prints,

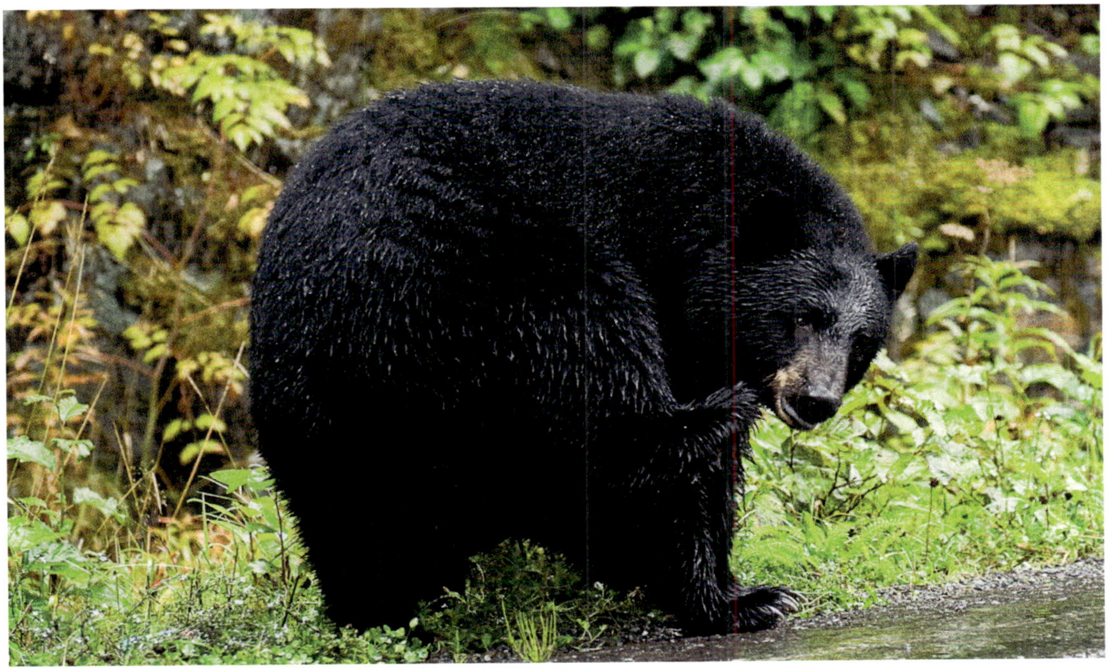

Figure 4-7. *Since the bear takes up most of the frame, vast numbers of pixels aren't needed. If the bear were 500 feet away, you'd need 35-40 megapixels to get the same quality.*

megapixels are not very important to this shot because so many of the available pixels are devoted to the subject of interest, i.e., Mr. Black.

Mamiya/P25 Nikon D700

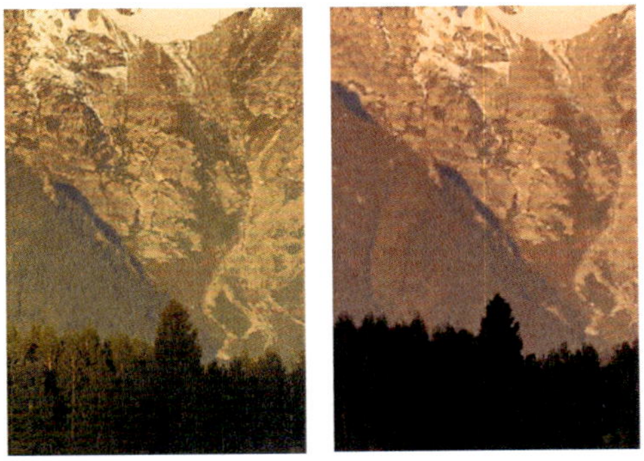

Figure 4-8. *Shows what resolution does for landscapes. Left image is 22 megapixels, right is 12 megapixels.*

The next two shots show the same scene shot with two different cameras, a Nikon D700 and a medium-format Mamiya AFD with a PhaseOne P25 back (22 megapixels). The Nikon had an f2.8 28-70 Sigma lens, and the Mamiya sported a similarly sized (though slower) lens. (An 80-millimeter Mamiya lens is equivalent to a 50mm Nikon lens in breadth of field). Both cameras were mounted on the same tripod and the shots were taken within a minute or two of each other.

The images differ some in color, but that's not the point here. The Mamiya camera / lens combo is FAR sharper and more detailed than the Nikon because 1. the glass is better and 2. there are more pixels devoted to the same portion of the subject.

Update

On February 7, 2012, Nikon released the D800 and D800e cameras. Cameras started reaching customers in late May. Nikon has apparently done the impossible – stuck 36 megapixels onto a 35 millimeter format sensor with extraordinary results, certainly comparable to cameras costing 4 to 10 times more! Here are sample shots showing just how sharp this camera is (100mm micro Nikkor lens helps).

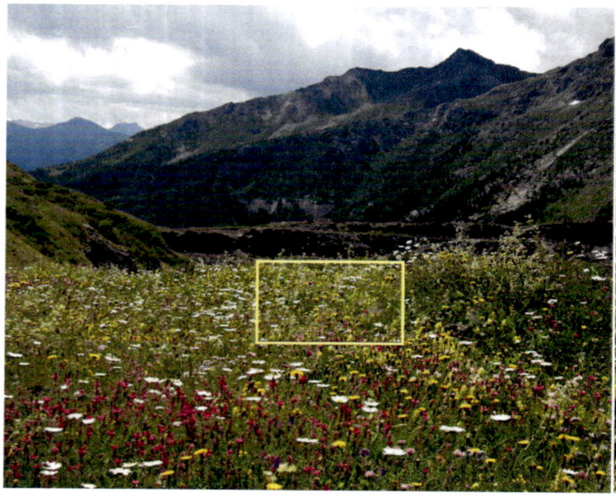

Figure 4-9. *Alpine Meadow*

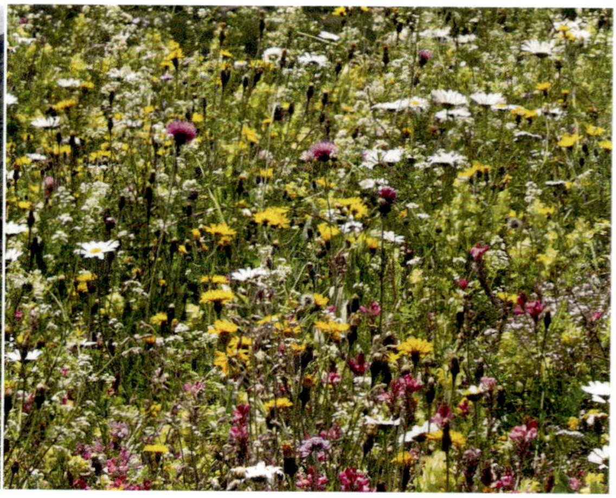

Figures 4-10. *100% crop of the area framed in yellow*

Big Pixels vs. Small Ones

Sorry, folks, size does count. Bigger pixels do a better job of capturing light than small ones. Imagine you are playing a carnival game in which you're tossing dimes at a tray crammed with red, green, and blue shot-glasses. You're aiming at the blue shot glasses. Some will go into the shot glass and win you a stuffed bear. But many will bounce off and either miss the shot glasses altogether, or land in the wrong one.

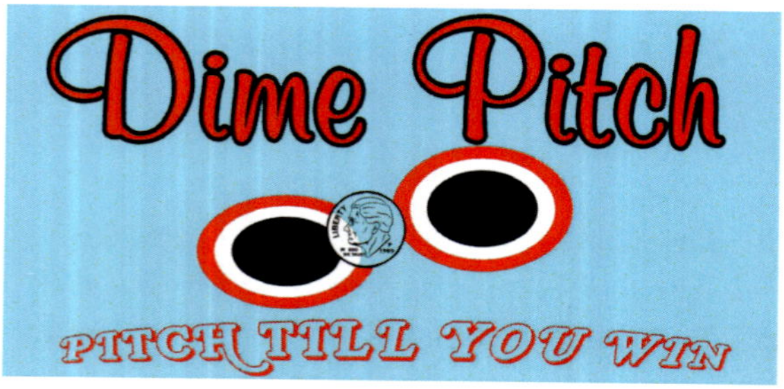

Figure 4-11. *Dime toss game*

The pixels on your sensor are exactly like the shot glasses – only the photons striking the bottom of the glass are captured. If the pixels are small and very tightly packed, many will scatter instead of going to the active part of the sensor and being recorded.

Now imagine that you're still tossing dimes, but this time the shot glasses have been replaced with coffee cups (no handles, so they're still tightly packed). This time most of the dimes will find a home (and the correct home at that). You win the teddy bear!

When you have a DX sensor (smaller than 35mm), and give it 10 megapixels, those pixels are mighty small and very tightly packed. That means that there's more light scatter, and the effect of that is less dynamic range. Bad. But if you take the same number of pixels and put them on an FX sensor, and if they're still tightly-packed, each pixel will be bigger and your dynamic range goes up.

On the Phase One P25 sensor (now obsolete) the pixels are 9 microns across. And there are 22 million of them spread over a sensor the size of a 645 digital negative. On a Nikon D300 the pixels are 5.45 microns across. On the D700, they're 8.45 microns across and there are 12.8 million of them on a 35 mm (FX) size sensor.

On the new Nikon D800, pixels are 4.88 microns across. There are 36 million plus of them.

So go back to the dime-tossing analogy. Bigger sensor+bigger pixels = more light captured. True — but in all honesty the D800 sensor has a dynamic range very close to a 40 mp Hasselblad medium-format camera and Nikon seem to have resolved issues regarding light scatter as described above.

Bit Depth (Color)

One more factor affects dynamic range, and with it, effective resolution (the amount of detail in shadow areas, for instance). This is bit depth. Bit depth is defined as the number of bits (binary digits) used to record the illumination level of a pixel. More is better. The D700 carries up to 14 bits of color information about each photon of light striking the sensor, while the P25 carries 16. This may not sound like much of a difference, but it's a very big deal. RAW image files from the Nikon are about 12 megabytes, from the P25 about 24 megs. Better have plenty of memory cards! Big files take more processing horsepower, and hence allow fewer shots-per-second. The P25 is dead slow (0.7 frames per second), while the D700 shoots up to 5 frames per second comfortably. Frames per second is irrelevant for landscape work, but vital for shooting sports or birds. Buy for what you need to shoot! The D800 yields files up to 45.5 mb when shooting in 14-bit color, smaller in 12 bit.

Drawbacks to Monster Cameras

When your are shooting with a camera like the D800 or medium format cameras, the sheer resolving power of the camera tests the photographer. Camera shake, inaccurate focus, and cheap lenses all cause flaws that are instantly visible at the huge print sizes native to these cameras. (The native print size of the D800/800e is slightly larger than 16x24 inches at 300 ppi). This means that both your equipment and your technique need to be first class. If they aren't, you will grow very, very frustrated.

Summary

The point of this article is not to compare specific cameras. It's to clarify when you need lots of pixels for maximum resolution and when you don't.

You need lots of pixels when you want to make razor-sharp wall-size pictures of landscapes OR when you want to make large prints of deeply-cropped subjects.

You don't need lots of pixels to shoot close-ups, or when you don't plan on printing large, or when you don't need to crop then print.

Most of us carry cameras with more megapixels than we need. But as it usually is in life, we make choices because some marketer tells us simplistic stuff like "More pixels is BETTER."

More may be better – but it depends on the jobs you want to do. Define what you're trying to do, then match the megapixels to the task – or contentedly waste your money for something you don't need and will probably never take full advantage of.

PIXELS PER INCH AND DOTS PER INCH (PPI AND DPI AND WHO CARES?)

Elsewhere I've written about resolution, primarily as it applies to sensor design. In this article I'm aiming at a less lofty goal … clarifying the difference between Pixels Per Inch (PPI) and Dots Per Inch (DPI). There are lots of explanations out there, but perhaps one more won't hurt.

Pixels are the smallest light receiving unit (on a sensor) or light emitting unit (on a monitor). Pixels come in various sizes and density. Camera sensors measure them in the millions, spread like salt (only very carefully arranged) on a sensor measured in millimeters. The pixels themselves may be of different sizes, the common range being 5.5 microns (millionths of a meter) to 9 microns. Since they're very tiny, you can pack millions of pixels (megapixels of 'em) onto a sensor.

Pixels on a monitor are different. They are commonly given in pixels-per-linear-inch. Despite what you used to hear about 72 per inch (mac) or 96 per inch (PC), most monitors have between 90 and 105 pixels per inch. If your screen actually has a pixel density of 100 per inch, that means you'll have 100x100= 10,000 in one square inch of screen. Compare that to the millions per square inch on your sensor.

Obviously pixels on a monitor are far more widely spaced and hundreds of times larger than the pixels on your sensor. The space between pixels on a monitor is called "pitch" and is commonly on the order of .27 or so millimeters between each pixel. I'll get to the so-what of all this in a moment; first let's look at DPI.

DPI – dots per inch – has nothing to do with sensors or monitors. It has everything to do with print quality. DPI is the number of dots of ink sprayed on the paper by an ink-jet printer. For normal print sizes and magazine use, 300 PPI is the common standard for image files you send to newspapers, magazines, and labs like Bay Printers or WHCC; below that, the image simply won't print well enough for magazine or book use when converted to DPI; art directors are just persnickety that way (see "A Lesson in Humility" in Chapter One).

So size your print file at 300 PPI and you're golden, right? Well, maybe not. First, very large prints do not need to be printed from 300 PPI files. I'm talking really large – 20 x 30 inches up to by 20 x 30 FEET. For instance, the graphics on airplanes are printed at 25 dpi.

Why? Two reasons. First, the bigger the print, the bigger the file needed to produce it. If you wanted to print a file for printing 20x30 feet at 300 ppi, you'd need a file measured in terabytes! (7.77 BILLION dots). And you still have to convert image data to printer data; watch out!

The second reason is that human eyes aren't that good. When you watch a TV across a room, it looks pretty sharp – even though the screen pixels are separated quite widely. That's because your eyes don't see well enough to see each screen pixel separately; we merge them together and form the image in our brain the way we think it OUGHT to look.
So billboards and airplane graphics don't need to be printed at 300 dpi or more to look sharp. If you're

seeing them from 150 feet away (and that's WAY closer than you want to see the tail of an airplane near you), they'll still look sharp at far lower density. So what if up close they're awful? Doesn't matter, and processing files does matter in terms of both time and equipment. So thank heavens for human limitations!

In fact, the further away an object is, the less clearly we can see it – thanks to atmospherics and some laws of physics and physiology that we don't need to mess with.

So, for ordinary purposes, you should try to size your image to print at your chosen size at 300 PPI, or no less than 250 (which will still look soft).

Wait a minute. Your camera and computer are seeing images in pixels per inch – but your printer is using dots per inch. How do you get from one to another?

The best explanation I've seen comes from Josh Lubbers, head of BlueCubit Software, and an old hand in the color wars.

> Customer has a 300 PPI image open in their application that they wish to print. They click Print and send the file to the print control panel for their Epson Printer.
>
> Printer Driver resizes the 300PPI image to 720PPI using nearest neighbor (low quality) or Bi-Liner (low quality) interpolation. This happens in 100% of all print jobs that are printed from Photoshop, Lightroom, etc. Why? An inkjet printer has a native resolution that all images must be set to in order to print. The same is true for Canon and HP printers, but their native PPI is 600 PPI.
>
> Print Driver applies a microweave process that converts the PPI (pixels per inch) to DPI (Dots Per Inch with options for 720, 1440, or 2880 DPI on an Epson printer- Canon and HP microweave to different sizes). Printers don't put pixels on paper; they print dots on paper and microweaving is the process of converting pixels to dots. There is no Image quality lost in this part of the process but a higher level of microweaving generally results in a higher quality print but a slower print time.

Why can't you just resize the image in Photoshop (or whatever) to the DPI of your printer? You can, BUT 1) you won't see a difference on your screen and 2) file sizes grow exponentially, and 3) the printer driver still has to convert to DPI.

So that's it. PPI are pixels, and are electronic. DPI are ink droplets, and are physical, not electronic. Your print driver takes your image and converts it to instructions to the printer to produce the appropriate DPI for printing.

I can't close without saying that Josh Lubbers makes ImageNest, a post-script RIP (raster image processor) for Mac OS only, that nests your images (to save paper) and resizes them. Version 3.5 bypasses the whole Epson or HP printer driver thing and uses a FAR superior method of up-rezzing (enlarging), and it also has a built-in sharpening routine that gets applied AFTER the image is sized for printing. As RIPs go it's inexpensive, and upgrades are always free. Yowzah!

CHAPTER 5: *Composition and Design*

THINKING PHOTOGRAPHICALLY
BASICS OF COMPOSITION
STORY-TELLING IN PHOTOGRAPHY

CHAPTER 5: *Composition and Design*

THINKING PHOTOGRAPHICALLY

Most of the time, I'm a pretty normal guy. But then, suddenly, driving along, taking a walk, visiting an office building, it comes over me ... some sort of mental switch is thrown, and I'm thinking photographically.

Sometimes there's a little forewarning. A few years ago, I was taking one of my daughters on a college tour. I felt this odd sort of itching, and said to Jenny, "I feel a photo coming on." Not two minutes later, I pulled off the interstate to take this shot:

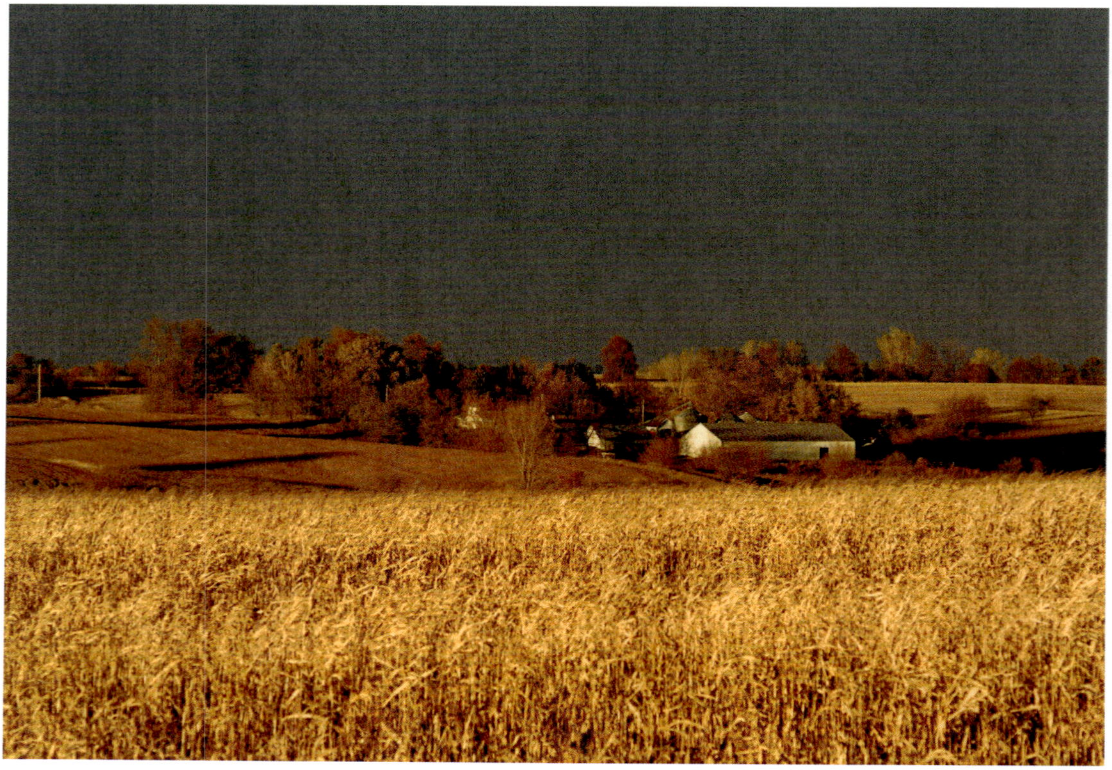

Figure 5-1. *Iowa farm*

Six minutes later, I stopped again and took this one, wholly different in mood:

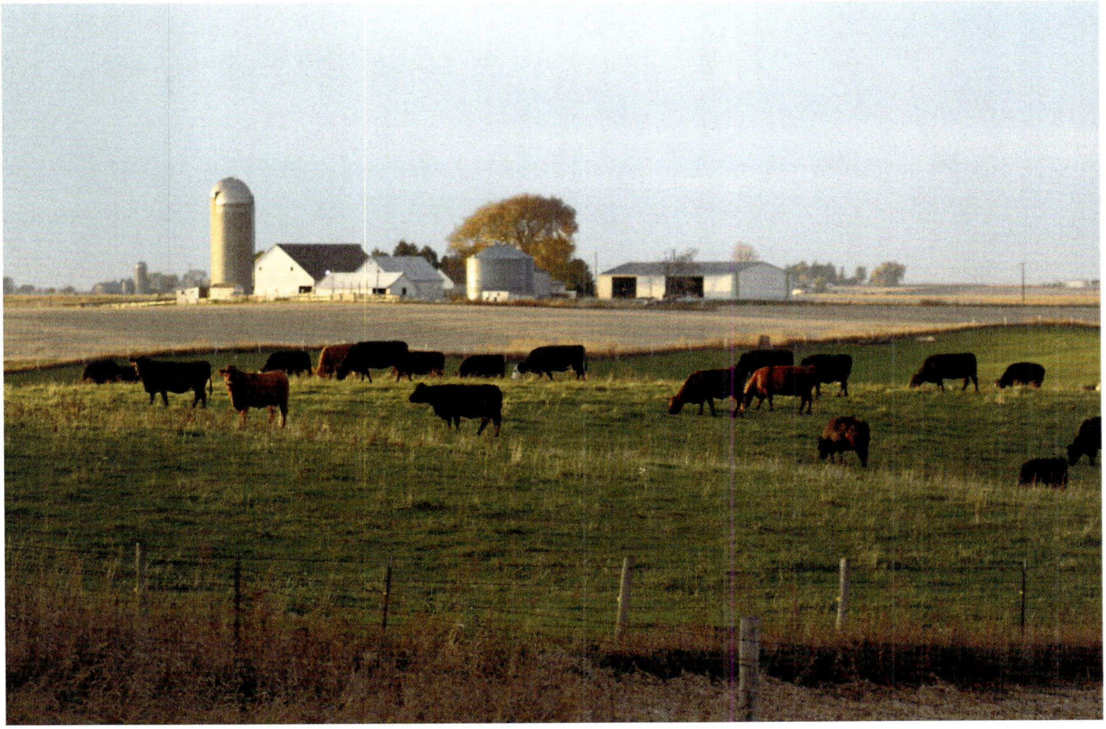

Figure 5-2. *Taken 6 miles down the road from 5-1.*

The point isn't that these are OK photographs; the point is that somehow I *knew* I would be taking them, because I had started thinking photographically. Perhaps you could call it *seeing* photographically; one's the cart and one's the horse, but I'm not sure which is which.

This article is an exploration of what it might mean to think photographically, what's involved, and how it can help you.

To me, thinking photographically means suddenly becoming aware of and attuned to the quality of the light, the shapes of the things around me (patterns, swooshes, patches of light and dark, interesting geometry, ironies, disjunctions, things that aren't looking the way you'd expect, or that are out of place, or that are picturesque in the old-fashioned sense of the word). I'm NOT necessarily thinking about barns and thunderstorms and late-fall corn stalks. In other words, I don't think it's so much the subject matter (though to some extent it must be) but the conditions that I'm suddenly aware of.

If you read the article on composition next in this book, these terms may make more sense. Suffice it to say that what I'm seeing when I think photographically is the only thing your camera can see: areas of light and dark, color, and how the areas of light and dark relate to one another. Your camera cannot see a barn. It sees areas of dark with white stripes and possibly a dark red color. We see a barn; it only sees light and the absence of light on a graduated scale (luminance).

When you think photographically, you think much more in the camera's terms than in your own interpretive terms.

Cameras are literal-minded, the Amelia Bedelias of the photographic world. They see light, they see color, they see dark, and they perform calculations and go click to create the "best" exposure. That's it. Humans are the opposite. We don't see light and dark, we see images. And not only do we see images, we see IMAGINARY images. Let me explain.

Everything we see we process conceptually as well as physically. That's how we know we're looking at a face or a snowman or a barn. We compare what we actually do see to what we've seen before, and bingo, that's a barn we're looking at.

We go further. We invent stuff that's not actually in the scene, and we suppress stuff that is in it. We may imagine we saw a cat in front of the barn, when the camera does not. We ignore telephone wires all the time, and we often don't notice the tree growing out of Uncle Peter's head because we are ignoring the background to concentrate on the foreground. Cameras don't see any part of the image as more important than any other part. Humans see selectively, and that's what makes us lousy eyewitnesses and sometimes gives us very nasty surprises when we get our pictures back from Wal-Mart.

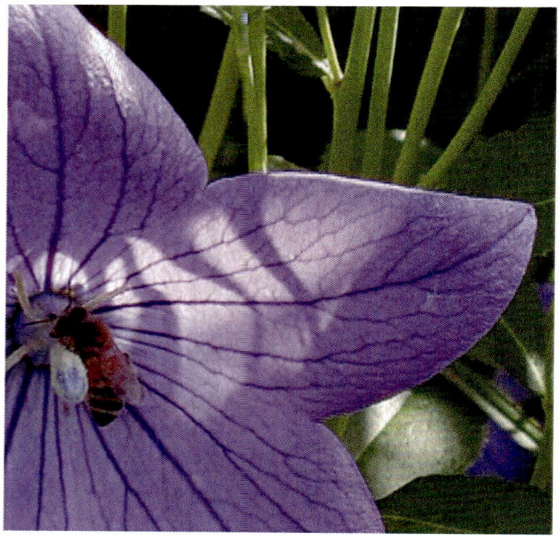

Figure 5-3. Morning glory with (invisible) bee.

Let me share an example. About 15 years ago I was taking a workshop from New England photographer Skip Schiel. Skip told us to go outside and shoot in the gardens, but NOT TO LOOK AT THE GARDENS, just the lights and darks. So I obligingly focused on a morning glory blossom, fascinated by the light playing across the petal. I shot very carefully. I never saw the bee that was gathering pollen until I got back to the computer and looked at the pictures. I was so into the task that I literally didn't see the insect right in front of my nose!

Oddly enough, this is evidence BOTH of thinking photographically and of human filtering. I was thinking about light and dark, but I suppressed something very, very visible – wasn't looking for it, so I didn't see it. (On a personal note, that workshop taught me a great deal and led directly to my getting serious about photography). Even when we're thinking photographically, we're still people, and we don't see very accurately.

Here is a trio of tips to help one think photographically.

- When framing a picture, stop looking at the subject and, one at a time, survey the corners of the frame. What's there that shouldn't be? Have you cut someone's arm off? Where did that tree branch come from?

- Look at the background through your viewfinder, and, as much as possible, only the background. Maybe you'll see the tree behind Uncle Peter, or the bull happily inseminating a cow behind your 6 year old twins! (Yes, this has happened.)

- When you see problems in your picture, MOVE either yourself or the subject. Often a few inches or a foot or two will take care of the problem, and you'll have a much stronger picture as a result.

In conclusion, cultivate the habit of thinking photographically. You don't have to wait until the fit is on you, you can go out and practice! Doing so will help you find more pictures, better pictures, and develop both your technical and artistic skills.

BASICS OF COMPOSITION

Composition, simply put, is the organization of visual elements in an image. WHAT??? You're supposed to ORGANIZE your photos?

Well, yes. Photos without composition or a clear organizing force (or hinge, or concept – whatever you want to call it), are confusing, chaotic, and dull. Even in close-ups and macros, where you wouldn't think much in the way of composition was available, it actually is.

For example:

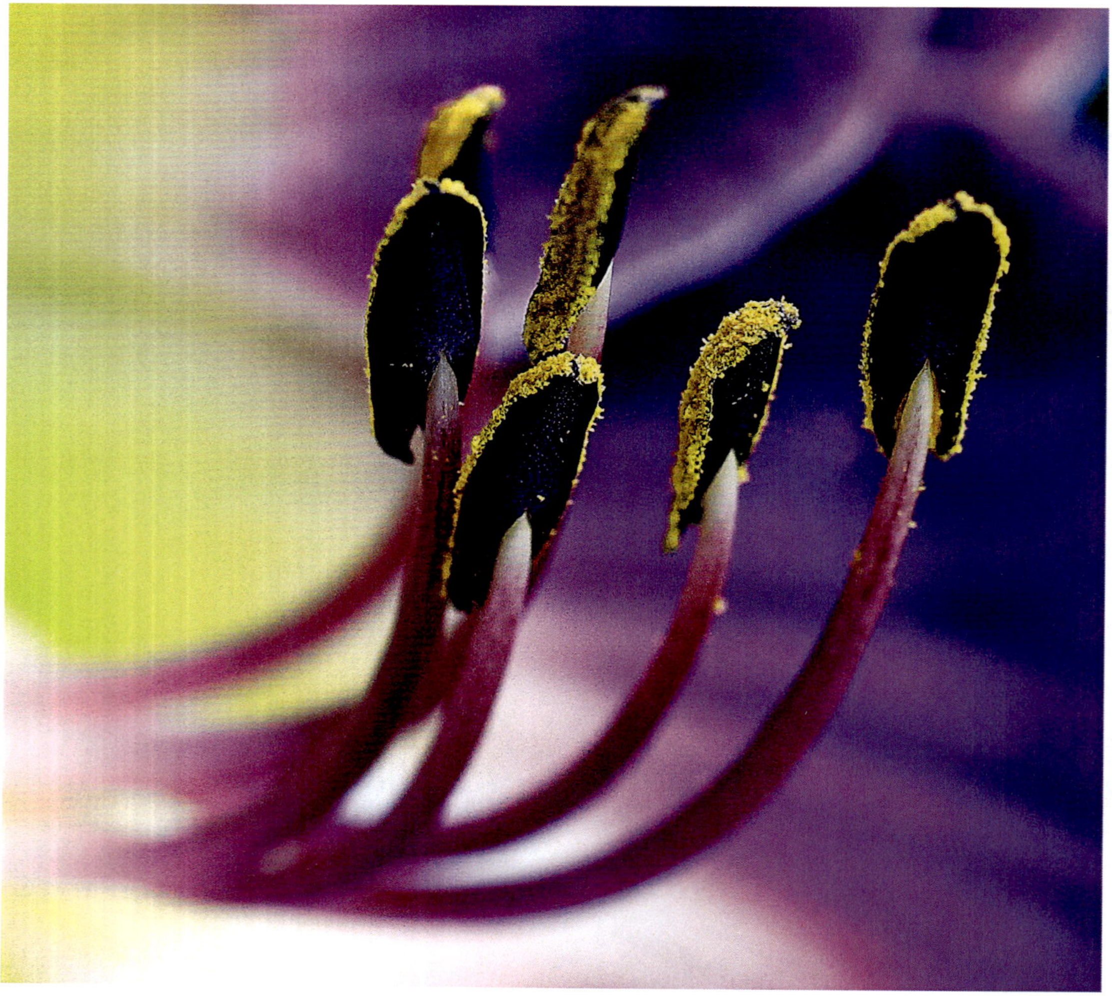

Figure 5-4. *Lily Stamens (macro)*

This picture is organized using a "swoosh" curve to energize the photo. The subject is isolated from the background by depth of field and the contrast of the bright gold pollen. Swooshes are wonderful, graceful curves. They are energetic but graceful, and far more effective at linking unrelated components than straight lines are.

The premise is that you want your pictures to be organized and energetic. Energy without organization is confusing; organization without energy is a bore.

What's energetic, and what's boring? Straight lines are boring. Boxes (at least one box by itself) are boring. Circles by themselves are boring. "Bulls-eye" composition – where the subject is in the dead center of the picture — is boring. Images without shadows to create relief and depth are boring. (This is called "flat lighting.")

What's REALLY energetic are curves, things that are a little off-balance, twists (remember the statue of the Discus Thrower?), dynamic light and shadow, diagonals, leading lines which pull your attention into or through the picture, and perspective (which leading lines help to create).

Things that look a bit unstable have a kind of pent-up energy that enlivens an entire picture.

As a small aside, you may wonder why the marvelous view you saw from the highway (and stopped to photograph) looks lousy when you get the print back or look at it in the computer.

There are a slew of reasons, but they start with

- No center of interest (no single thing that stands out)
- No perspective
- No depth
- Flat lighting

When you stopped your car to take the picture, it was memorable to you not for its bare-bones content, but because the sense of space excited you, or the air was exceptionally fresh, or it meant freedom to you, or some other reason the camera cannot photograph. Signs by the road with a camera silhouette rarely live up to their billing

Just flip this list over to see what your photos need: a single center of interest (subject), perspective, a sense of depth, and lighting that uses both light and shadow to bring your subject to life.

OK, so how do you achieve these things? First is eye training. Learn to see patterns as well as individual objects, or even instead of individual objects. Look for geometry. Look for fractals. Look for interrupted patterns of light and shade. Look for repeated shapes.

You can do this by looking at your own photo files. Pick out your best photos and see if you can discover how they are organized. You can even cover the subject of the photo and see what's left to hold the image together.

Grab your camera and go to a city or even village. Take as many pictures of patterns and shapes as you can find.... Don't take whole big scenes, but concentrate on things like parking meters going off into the distance, or rows of houses, or railroad tracks. These are all things where repeated patterns stand out. Do this over and over again for a couple of weeks. You'll be sick of it, but you'll start seeing the compositional "glue" that holds pictures together.

Here are some photos in boring and non-boring versions. What's the difference?

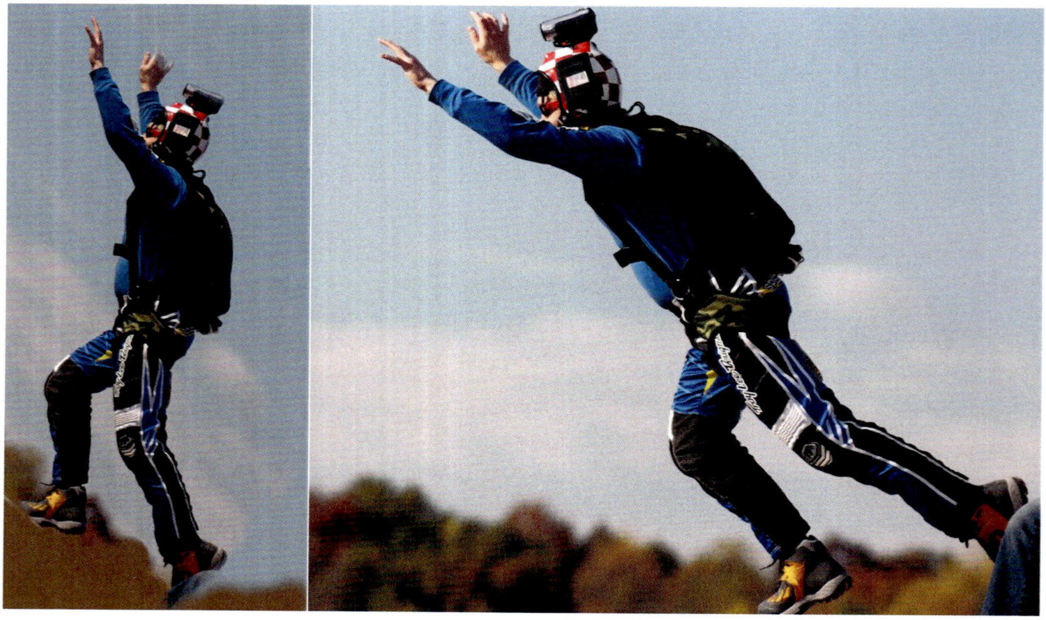

Figures 5-5 and 5-6. *Figure 5-6 shows the power of a diagonal in energizing photos.*

Yes, I've rather sloppily filled in missing areas on the top and left of Figure 5-5, but clearly, the diagonal is the visual element that makes Figure 5-6 so much more exciting than Figure 5-5.

Here's another pair. Figure 5-7 shows "bulls-eye composition." Yawn. Figure 5-8 is much better. Same photo, different composition. The second version is not cropped at all.

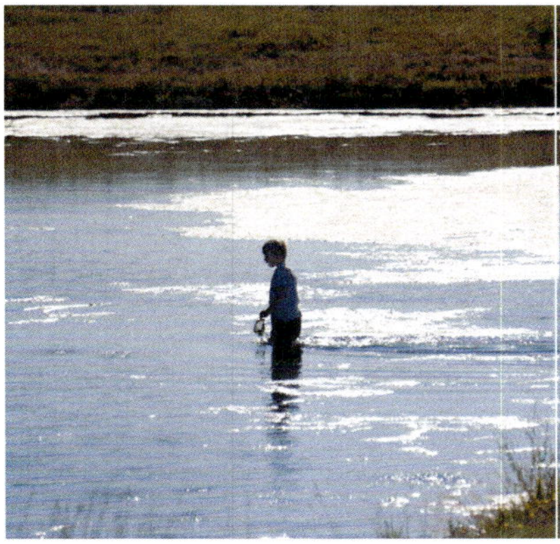

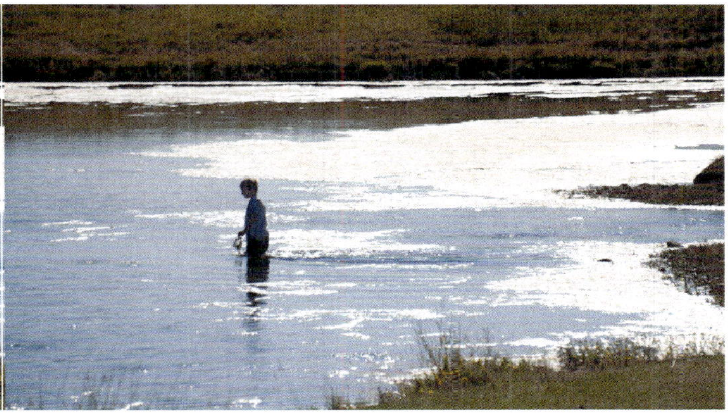

Figure 5-7. *Uses "bulls eye" composition.* ***Figure 5-8.*** *Relies on an assymetrical composition to energize the frame.*

The uncropped photo is far more interesting, partly because more of the background is shown, but mainly because of the position of the boy in the frame.

How do you know where to put the boy in the frame? Here is a tried and true way. It's called the "Rule of Thirds." The Rule of Thirds is derived from geometrical relationships that have been known since the Greeks. The grossly simplified version used by most photographers works this way:

Imagine a tic-tac-toe grid superimposed over your image. Each line of the grid is one third of the way in from the edge. The nodes where the corners cross are the highest-energy (and most satisfying) points in the picture. But putting your subject along one of these imaginary lines works, too.

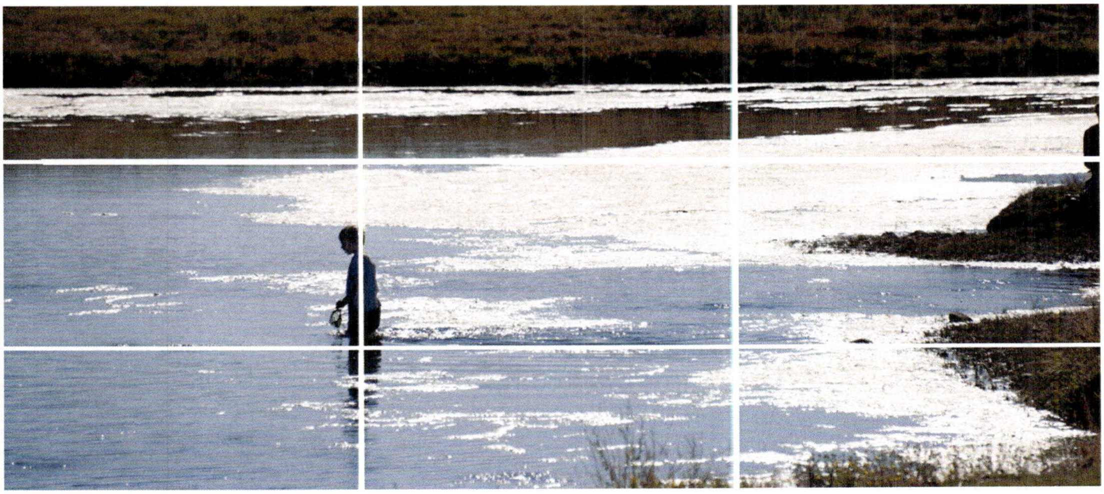

Figure 5-9. *Illustrates the Rule of Thirds.*

Using the Rule of Thirds is a great way to observe the basic principle: make your subject the center of interest, but not necessarily by putting it in the center of the picture! Other ways to do that include making it brighter than the background, or at least contrasting with the background, making it significantly larger or smaller than other objects in the frame, and isolating it in the frame. This photo illustrates all but one of those techniques.

The Rule of Thirds, useful though it is, is not the only way to organize your photos. Here are some others.

Diagonals

Diagonals impart LOTS of energy to your image. Even implied diagonals (such as eye contact between mom and kid) will do the trick. The parachute jumper above shows the principle at work.

The picture below relies on diagonals, but is a little more complicated. It has lots of power, largely because it bears the Mark of Zorro, created by contrasting light and dark areas in the photo. It's worth noting that the curve in the diagonal of the Z really helps; curves are never static while lines usually are.

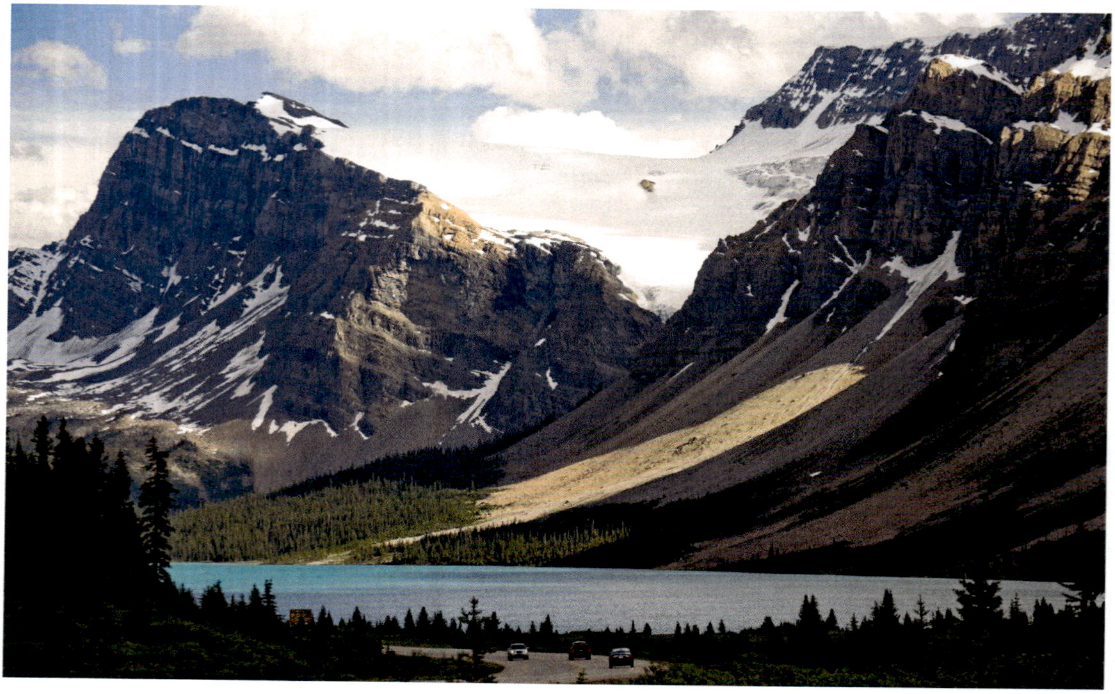

Figure 5-10. *Leading the eye along lines of contrast*

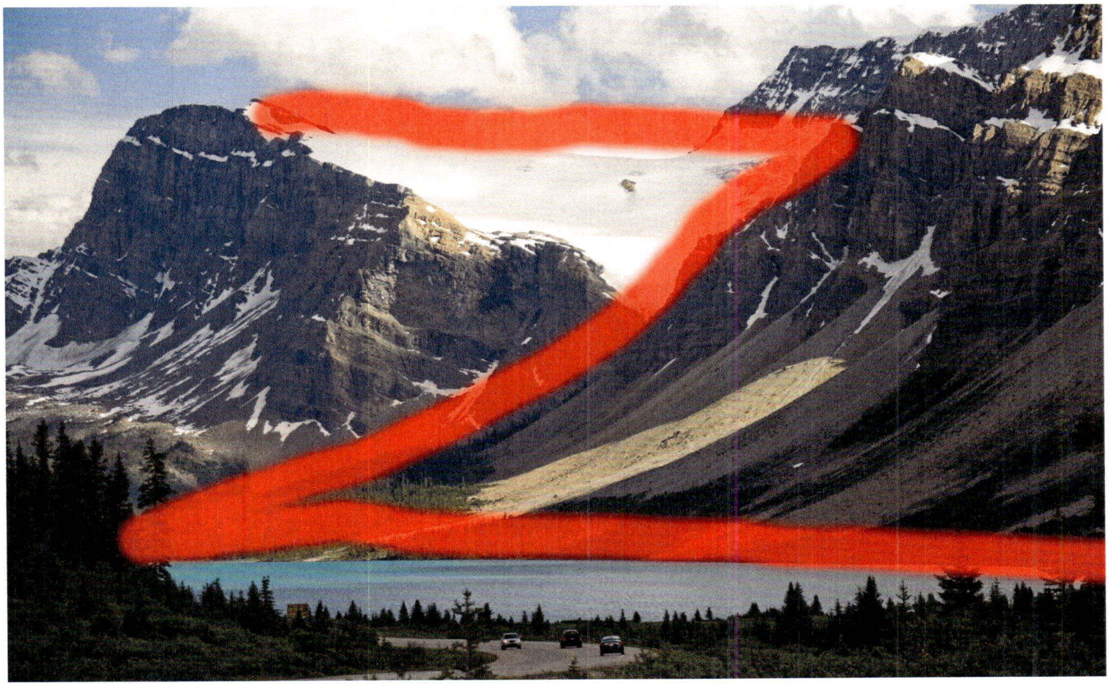

Figure 5-11. *The Mark of Zorro! See it now?*

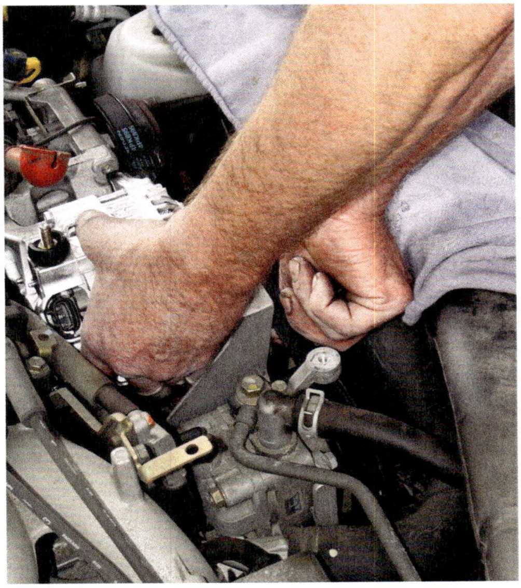

Powerful diagonals aren't just for landscapes. Here's one where the mechanic's left arm says volumes. (And my thanks for him working on a Saturday to replace an alternator in Beckley, WV). Note that the bright surface of the alternator makes s suitable end point for the diagonal.

Figure 5-12. *Strong diagonals lead the eye to the hands. The bright spot of the alternator also beckons the eye to what's important in the picture.*

Swoosh

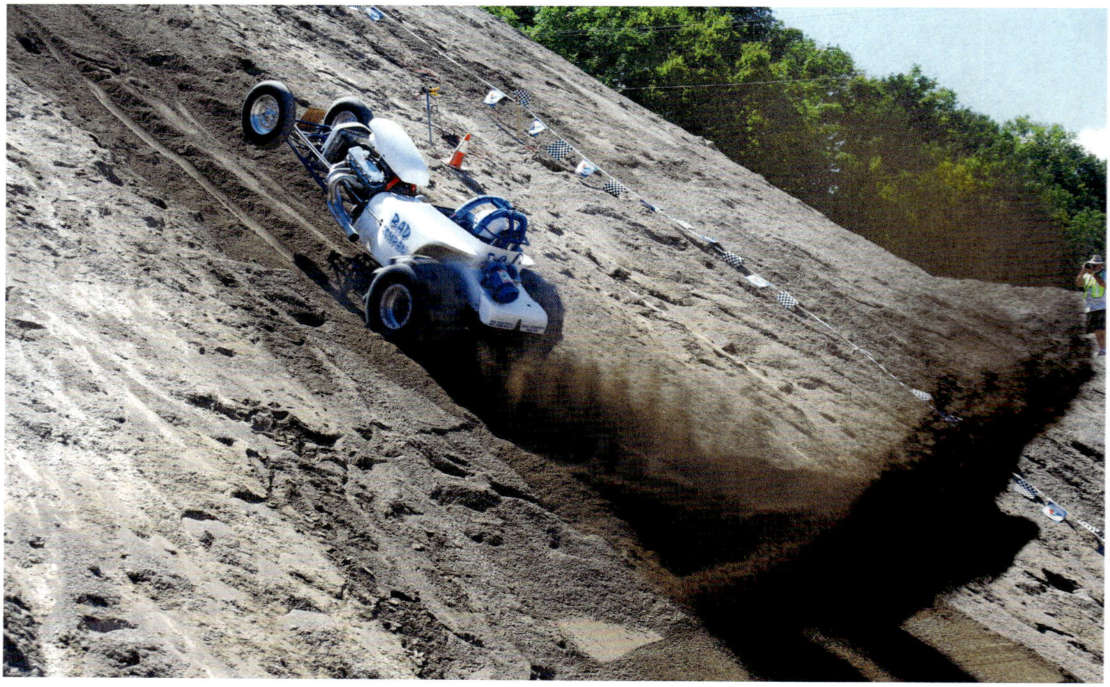

Figure 5-13. *The "swoosh" is another powerful compositional 'hook"*

Swoosh composition does just what you think it does – use a Nike-esque shape to give grace, power, and order to your photos. The swoosh may be subtle or obvious, but it is extremely effective. Relatively few photographers knowingly use this design element, but it works great and is well worth incorporating into your own work.

Repeating forms or patterns

Figure 5-14 uses repeating rectangles formed by a fencepost, the wire fence, and the field behind it, to organize the scene.

Figure 5-15 uses a combination of compositional elements to increase its impact. It features a bright subject, the Rule of Thirds (approximately) with an important plus: what are called "leading lines." Leading lines are just that—they lead the eye to what the artist wishes you to look at. In this case the fence stanchions are repeated forms, while the shadows and horizontals create perspective AND leading lines bring the eye to the walker. Also, the white tee shirt stands out and identifies the center of interest. Yowza!

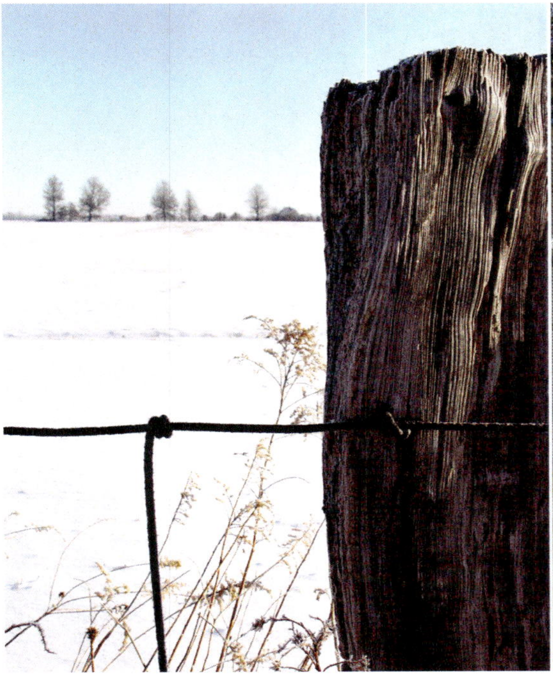

Figure 5-14. *A winter field and fencepost*

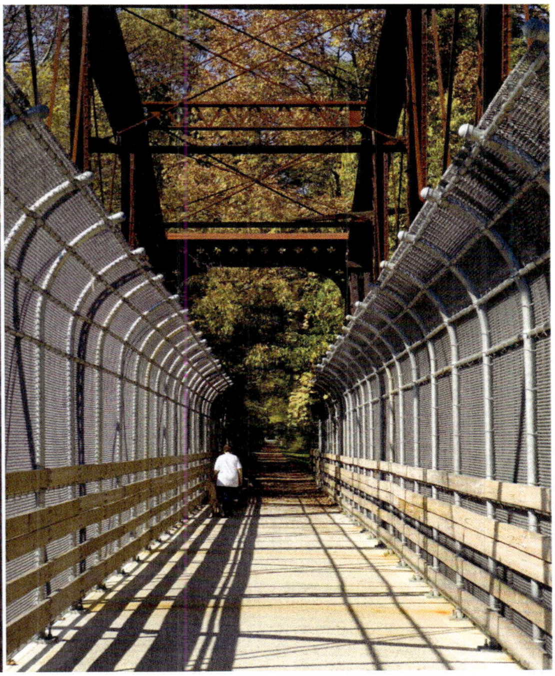

Figure 5-15. *Bike path bridge in autumn*

Figure 5-16 provides another example of repeated forms, plus strong textural and light contrasts to create counterbalances in the picture. Note the diagonal shadow extending from lower left to upper right, pulling the eye along that line and emphasizing the idea of division in the picture.

You're actually looking at an immovable object (rock) meeting an irresistible force (glacier) and the stress cracks and pressure waves resulting from the collision. It's a battle between titans, and all the oppositions in the image reinforce that idea. This one belongs both in a geology text and on your living room wall. Thank you, composition!

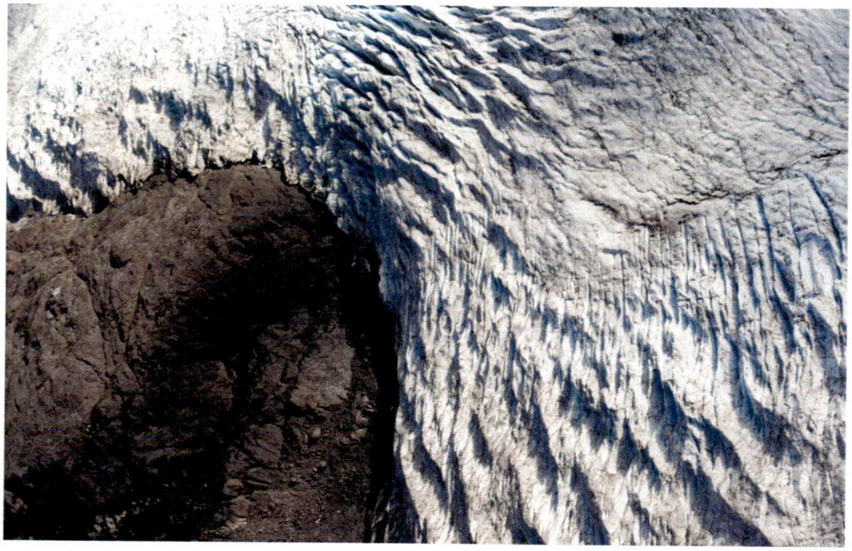

Figure 5-16. *Uses repeated forms, curves, and oppositions to make its effect*

Serpentine or S-shaped

Another really useful compositional scheme is the serpentine, or S-shape. S-shapes lead the eye into the distance and they're restless, never still. Result: an energetic shot.

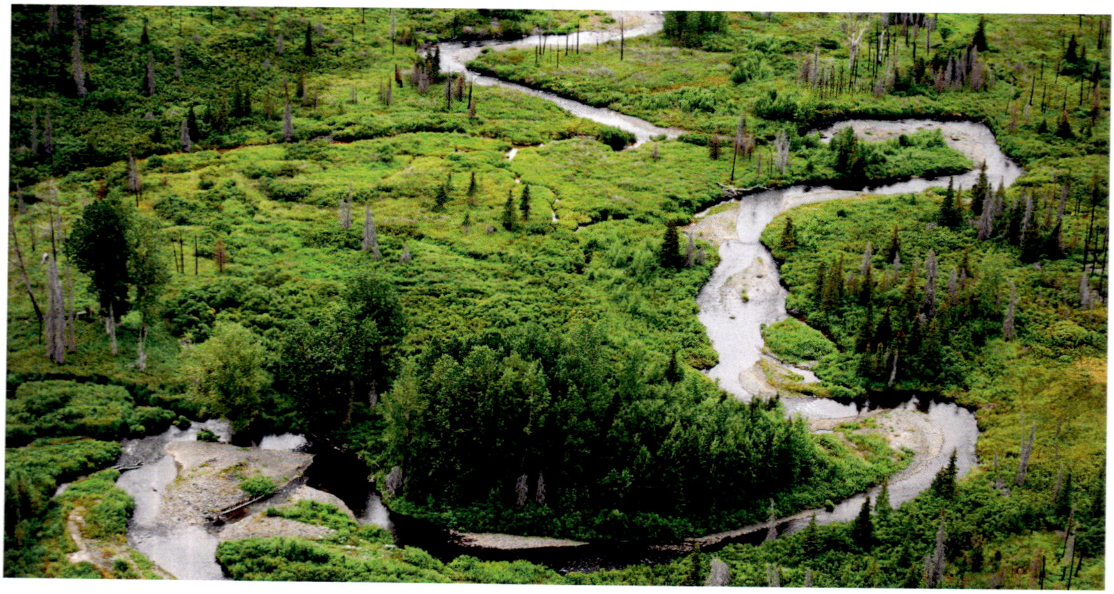

Figure 5-17. *Serpentine stream snakes through Alaska Tundra*

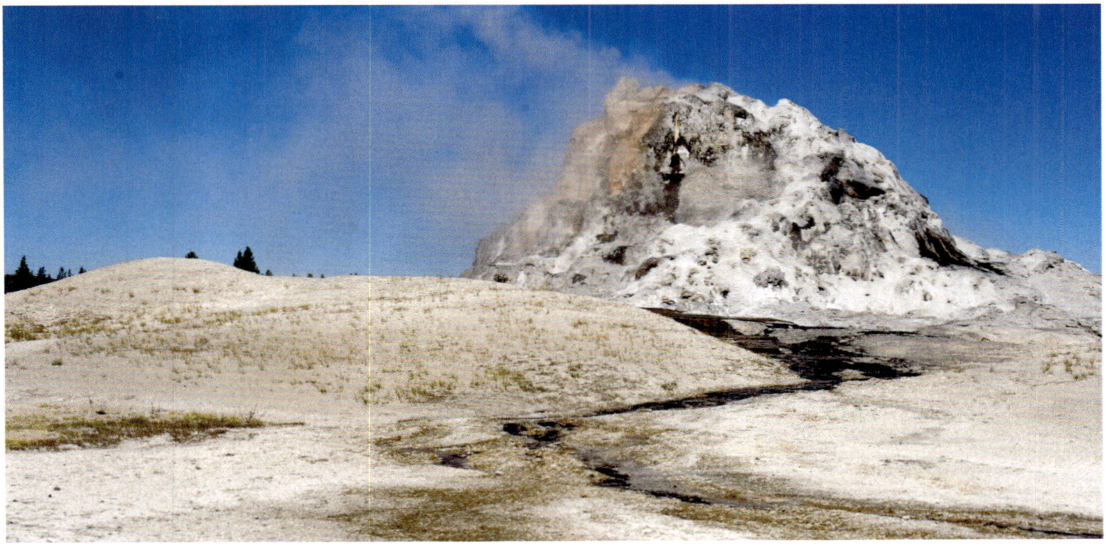

Figure 5-18. *Another snake slithers into the background. Even the smoke contributes.*

Test Yourself

You're probably getting the idea by now. Here are two mini-quiz photos. How are they organized? Which principles to they use? Figure 5-19 shows a palisade near Jasper, Alberta. What are the organizing principles in this picture? What are the things that make it energetic? What would you change to improve it? Hint: think light and dark contrast.

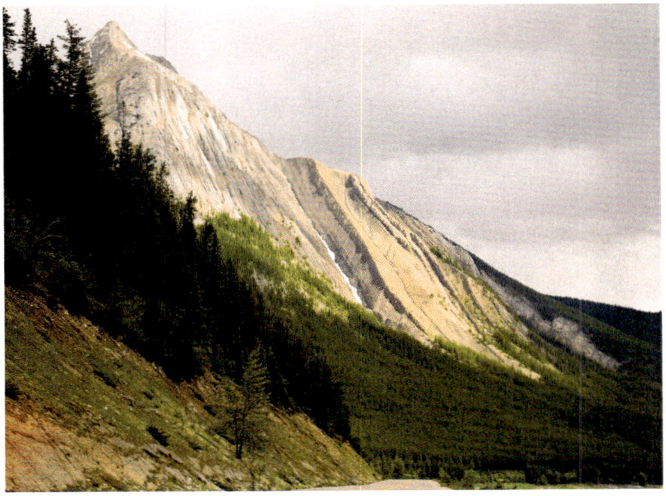

Figure 5-19. *Palisades near Jasper, AB*

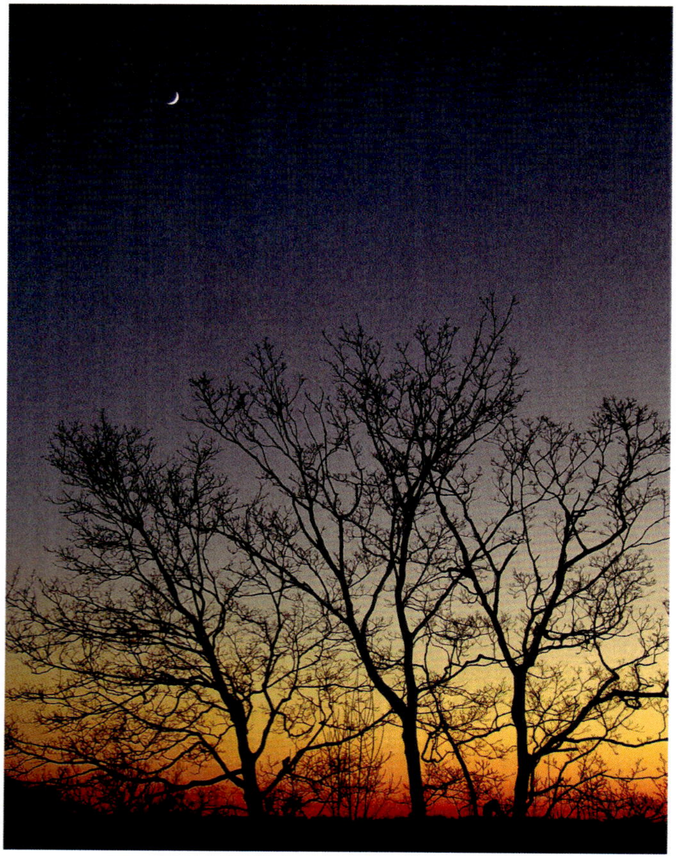

Figure 5-20. *Crescent moon, winter sunset, Loveland OH*

Hint on Figure 5-20: cover the moon and see what happens. What does the moon do, in terms of composition?

Summary

Photographs need to be organized and energetic. Composition is the key to achieving these qualities. Composition is first a matter of how you see, how you look at things, and secondly a deliberate (or instinctual, if you're really good) use of organizing techniques and principles to achieve results.

STORY-TELLING IN PHOTOGRAPHY

Story photos tell stories. Well, duh. But learning HOW story photos work, and what the elements are, may help you become a better photographer.

Story-telling is one of the things that make photos particularly strong and effective. According to the Professional Photographers of America (PPA), story telling is one of the 12 criteria for an excellent print. Unfortunately, they put it last, and I think it should be number 2 or 3. What PPA doesn't do is define what story-telling is and why it's important. Here's what they do say: "Story Telling refers to the image's ability to evoke imagination. One beautiful thing about art is that each viewer might collect his own message or read her own story in an image."

That's it. No wonder they have trouble judging this element in print competitions.

This article is intended to explore what story-telling in an image means, and how it comes to be. Additionally, a number of tips are provided to enhance your ability to create story-telling pictures.

What's in a story-telling photo?

- Subject (usually characters, whether people or animals or symbols to which we can relate)
- Situation (setting, place, props, clothing, time of day, etc.)
- Events
- Relationship (plot)
- Time
- Technique

Story-telling photos engage the viewer in any of several ways. First, something in the image's content or subject brings about an immediate emotional reaction. The situation and/or events and/or relationships speak to us. We can relate to the situation, and we react to it. That reaction may be shock, surprise, fright, laughter, love, suspense, wistfulness, sadness, curiosity, puzzlement, delight, or almost any other feeling you can name. If there's no reaction, you're not looking at a successful story-telling photo.

Photo stories are about a subject —usually a character, or more than one. Photo stories are about relationships, either between / amongst the subjects, or between the viewer and the subject. For instance, a photo of a bunny by itself is not a story. But put the snout of a coyote in the background, and now you have a subject and a relationship. Wiley and the Bunny are in a relationship, no doubt about it. Sometimes the relationship is between the subjects in the shot, sometimes it's with the viewer, based on the situation. But relationship is one of the things that make story photos work. I'll come back to this topic later in this article.

Here's an example of a strong story-telling picture:

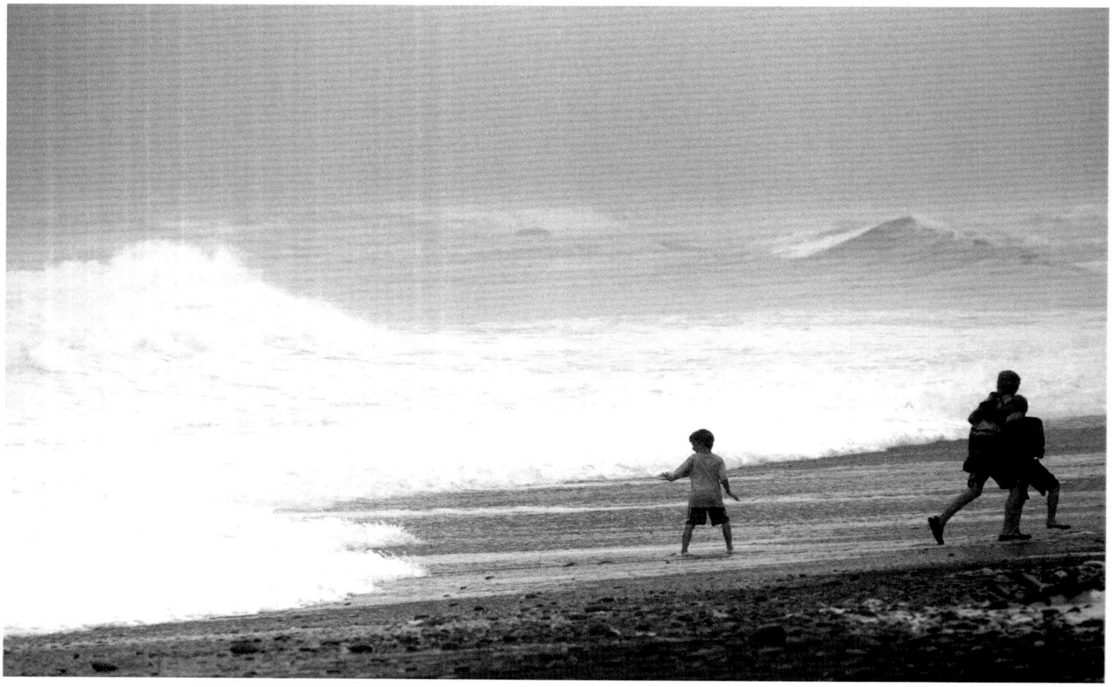

Figure 5-21. *King Canute Jr holds back the tide*

In this picture we see a complicated story. First, we see the waves rushing in, the boy daring them to swamp him, and his older brothers running for the high ground. That's the situation.

What makes this a charming story is the little boy's courage in facing the waves while the older boys hightail it. We see a lot of the boy's character in this shot, and we see the relationship with the older boys. He's the younger brother, but he's braver than they are!

This picture creates suspense but does not resolve it. How does this story end? Is the boy swept off his feet? Do the brothers rescue him? How long can he hold out? We are invited to make up any ending we like! That's engaging our imaginations, as per PPA.

You can say this story is about the boy and his reaction to his physical environment (the waves) and his social environment (his relationship with the brothers). Either way, it's rich in possibilities.

Story is good, but having plot in your pictures is even better.

In *Aspects of the Novel*, E.M. Forster explained story this way (I'm paraphrasing):

Story is events in their order. Example: The king died, and then the queen died.
Plot is different. Example: The king died, and then the queen died of grief.

Plot comes into being when you add relationships, causes, and consequences to events.

Good story-telling pictures depict events; great ones suggest a plot. And of course, the photographer has only his eye and shooting skills to bring that plot into being.

Here's a really famous story picture:

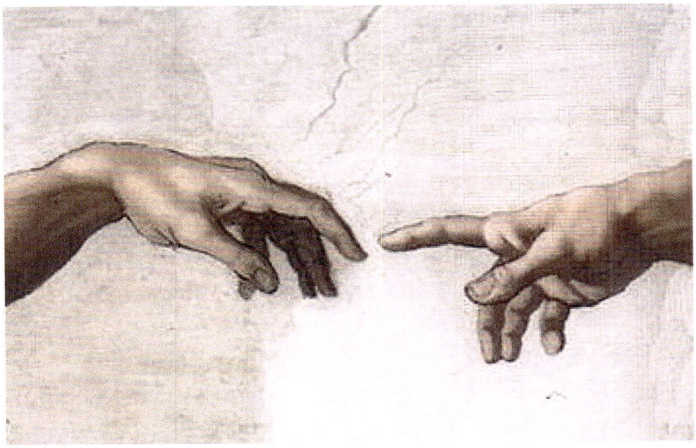

Figure 5-22. *Ceiling of the Sistine Chapel, Vatican City (not a photo)*

This famous detail from the ceiling of the Sistine Chapel shows the instant before God invests Adam with the spirit of life. You can tell who is who; God's hand has energetic tension and Adam is limp-wristed. The close proximity of the two fingers that you just know are about to touch creates the story. So we have characters, impending action, and we're all hanging there waiting for it to happen. There's lots of suspense in the story–even though we know how it's going to end.

By the way, hands make a wonderful photographic subject for story-telling pictures because they are so expressive of everything it means to be human. Here are four small examples from my files. Each uses props to establish the situation and body language to tell the story. What's the story each photograph tells? Or is it actually a story-telling picture at all?

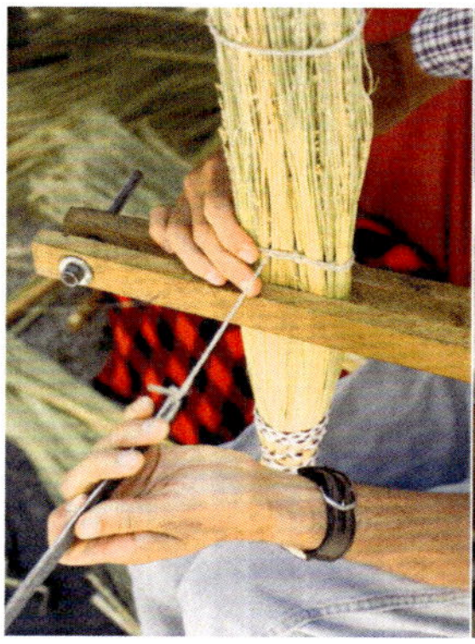

Figure 5-23.
Broom Making

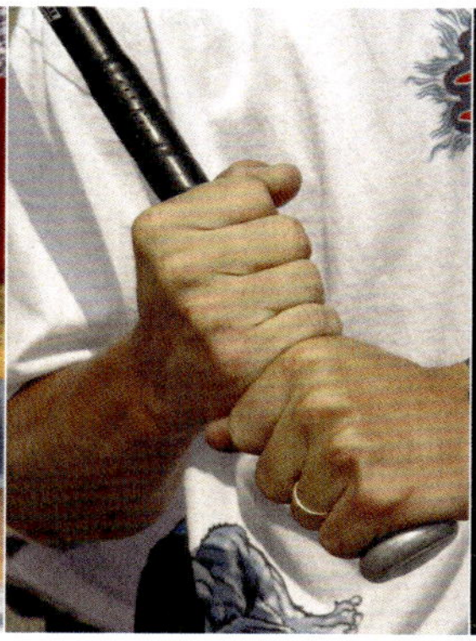

Figure 5-24.
At Bat

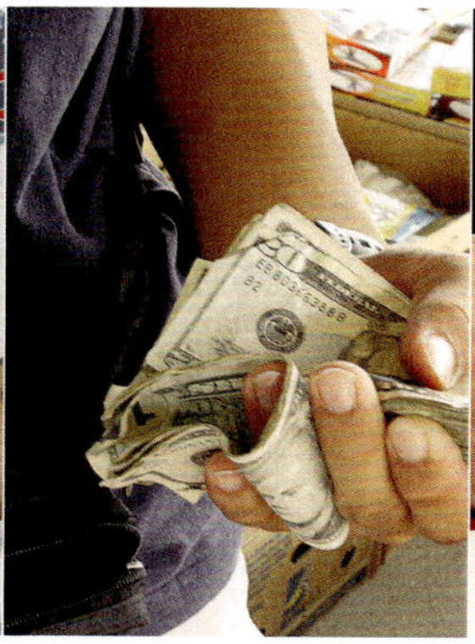

Figure 5-25.
Fist Full of Dollars

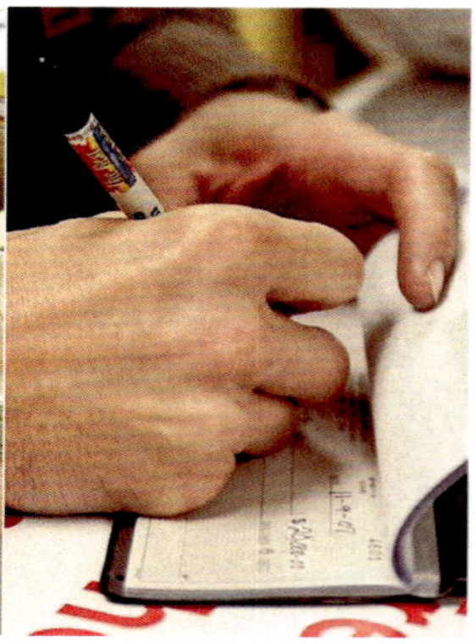

Figure 5-26.
Signing the Check

Time and Events

An event is an action, and it takes place in a moment of time. The way time is handled in still photography is tricky. You can capture the start of an event, the middle of it, or the moment it ends (or just after that). But you can only capture ONE moment per photograph. That's the difference between stills and movies, and it creates both a challenge and an opportunity.

The challenge is to pick the moment that has the maximum impact OR maximum suspense (poised between beginning and outcome). The opportunity is to engage the viewer's imagination as completely as possible. It's obvious that since a photo can't show the actual passage of time, it invites the viewer to fill in the blanks – again engaging the imagination, as per PPA. Story-telling photos engage the imagination by setting up a scene so that viewers fill in the missing pieces for themselves.

Here are six photos stressing time. What is the story in each one? Is the shot a beginning, middle or end? What do you think the outcome will be in each case? What role does time play in the story each tells? (I hope you'll take the time to answer these questions aloud).

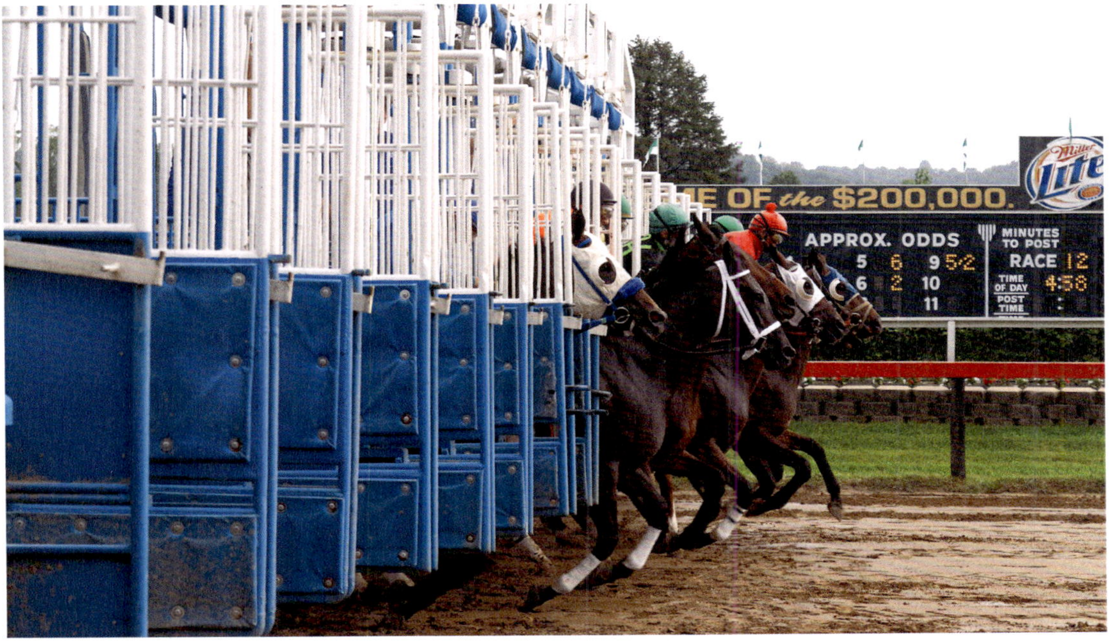

Figure 5-27. *They're off!*

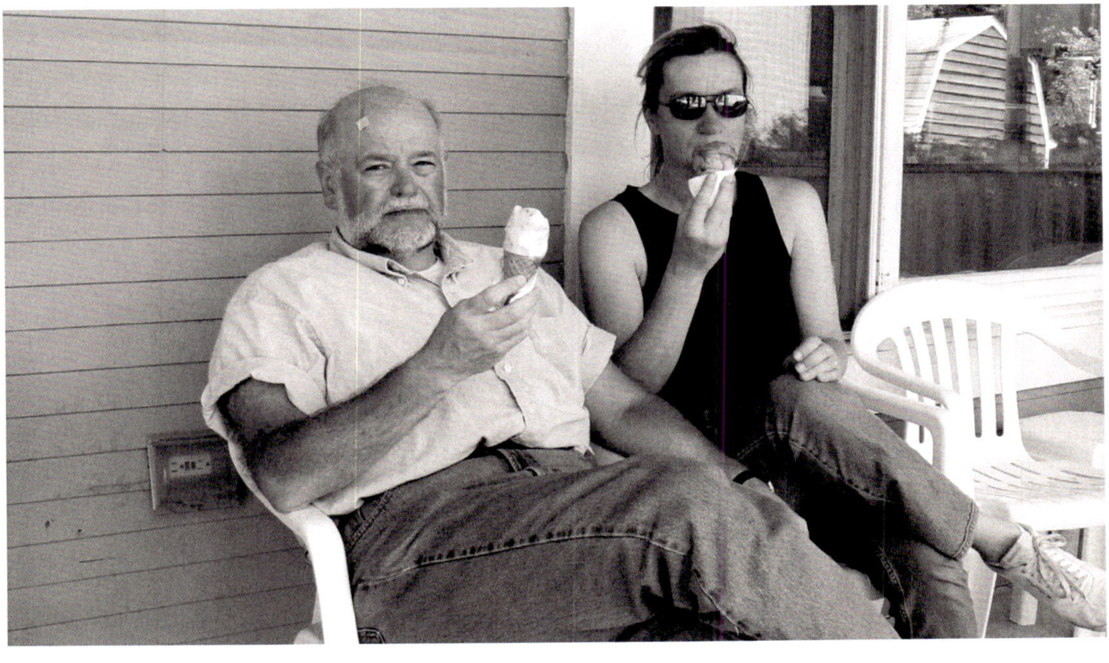

Figure 5-28. *Ice cream on the Fourth of July*

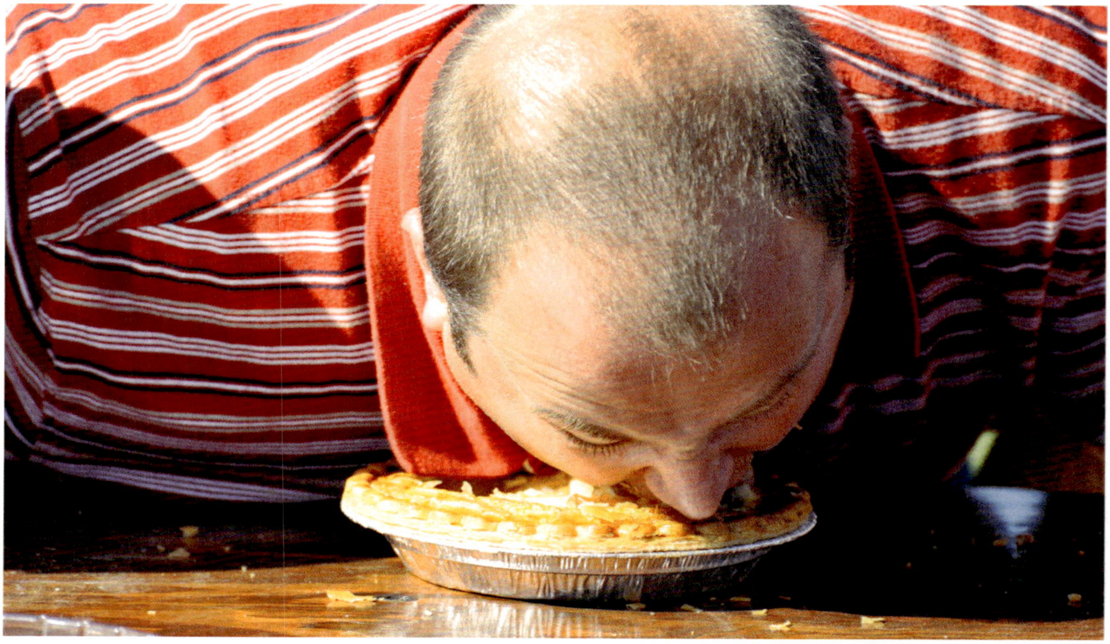

Figure 5-29. *Pie Eating Contest, Wilmington OH*

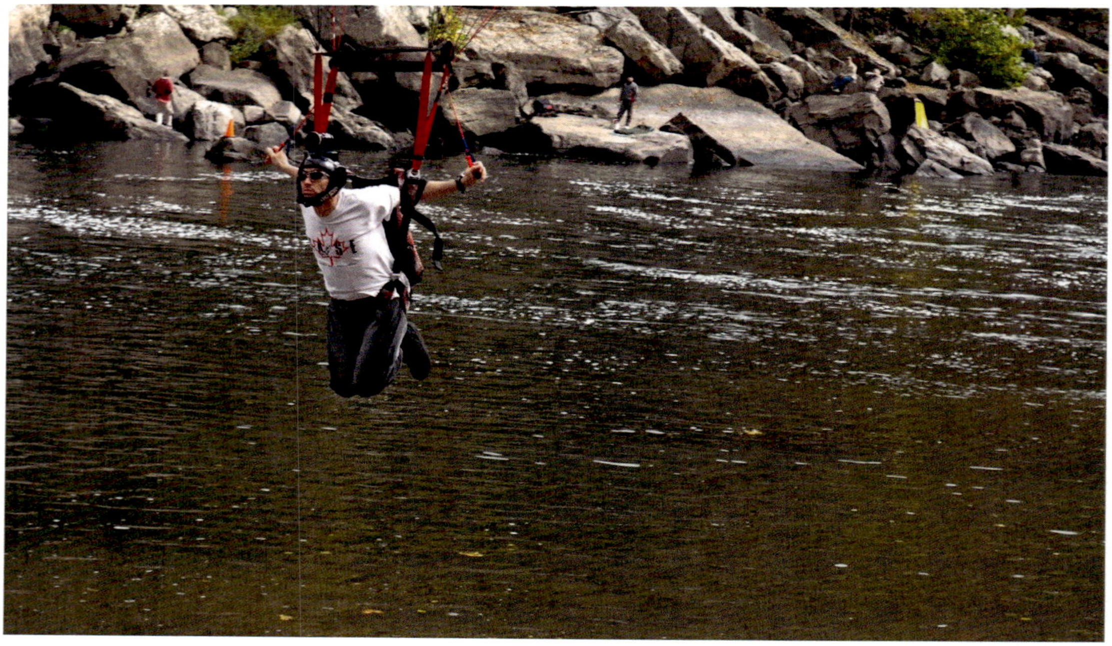

Figure 5-30. *C'mon, c'mon. c'mon!*

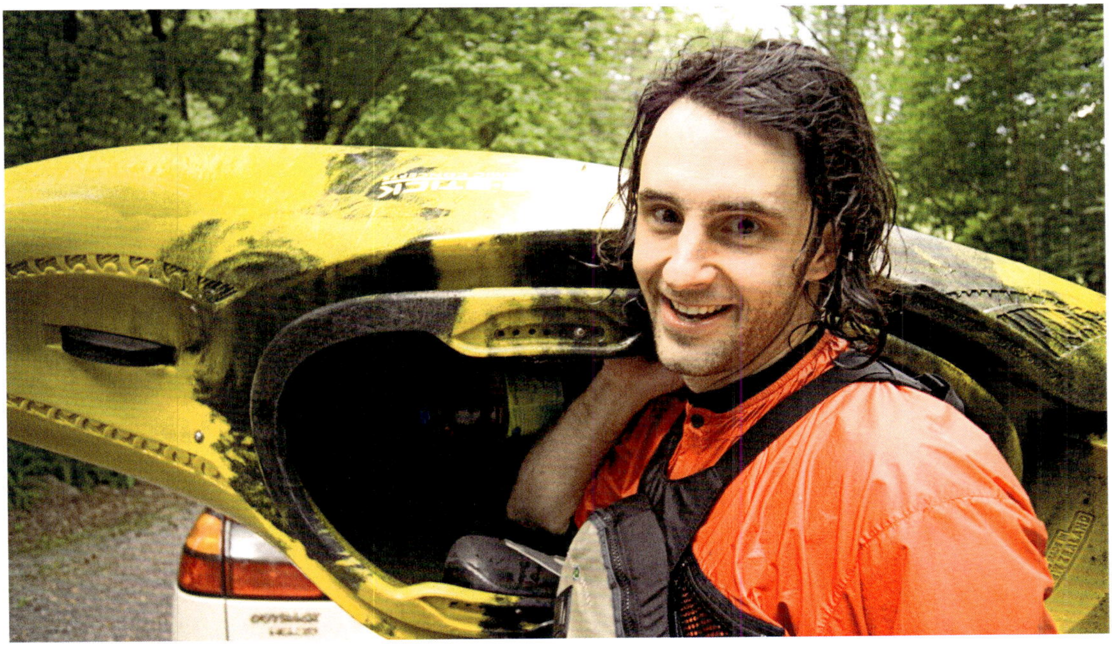

Figure 5-31 *That Was Fun!*

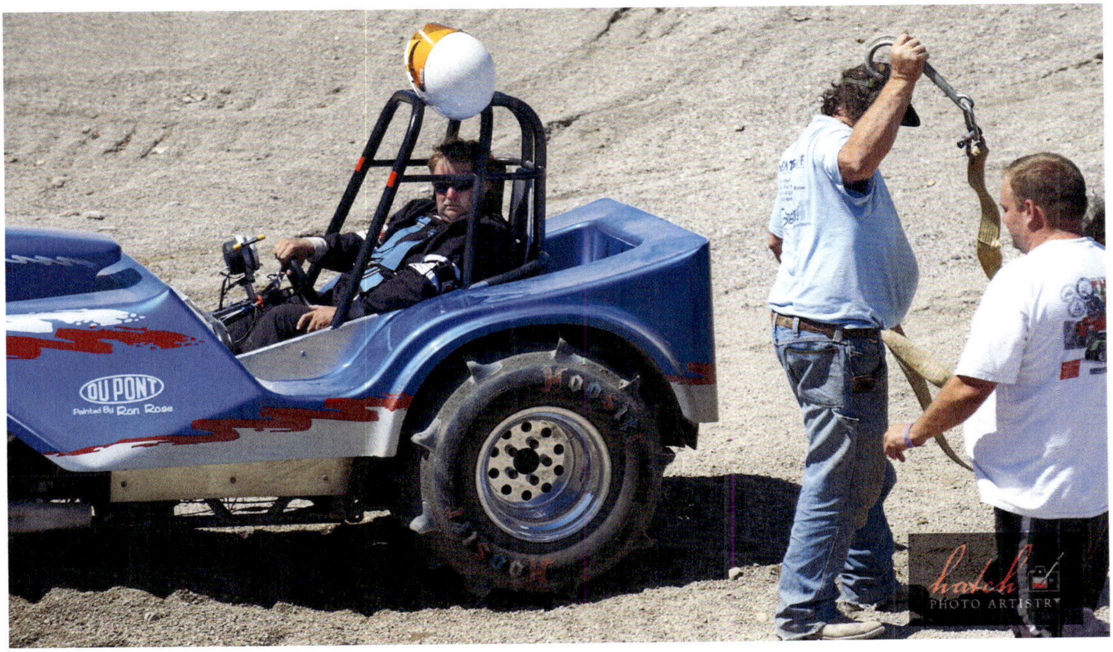

Figure 5-32. *Damn! Gravelrama Hill Climb, 2011, Cleves OH*

Stories and Relationships (again)

Stories are usually about people – and especially about relationships. But not every "people picture" is a story picture, and a close-up of a smooching couple may NOT work as a story picture (no event, no setting). A picture of a West Virginia coal miner sitting on a porch drinking a beer may be a fine picture, though not much story. But take that same man, photograph him wearing his lamp, streaked with coal dust, and emerging exhausted from the pit-head – now you have both a portrait AND a story.

The difference between the story and non-story photos of the same man is in the clues about him provided by the environment: props, costuming, lighting, composition, color (if any), and expression. French film makers call this quality "mis-en-scene."

In short, there's a relationship between the subject and the environment in any story-telling picture.

The second great element is, of course, relationships among people. A photographer captures these relationships using the same elements of timing and mis-en-scene mentioned above. How do you do it? One great way is to emphasize "expressive" body parts – faces, eyes, mouths, hands, or even feet. Eye contact between the subjects can create a link stronger than an iron bar. Take this casual photo for example:

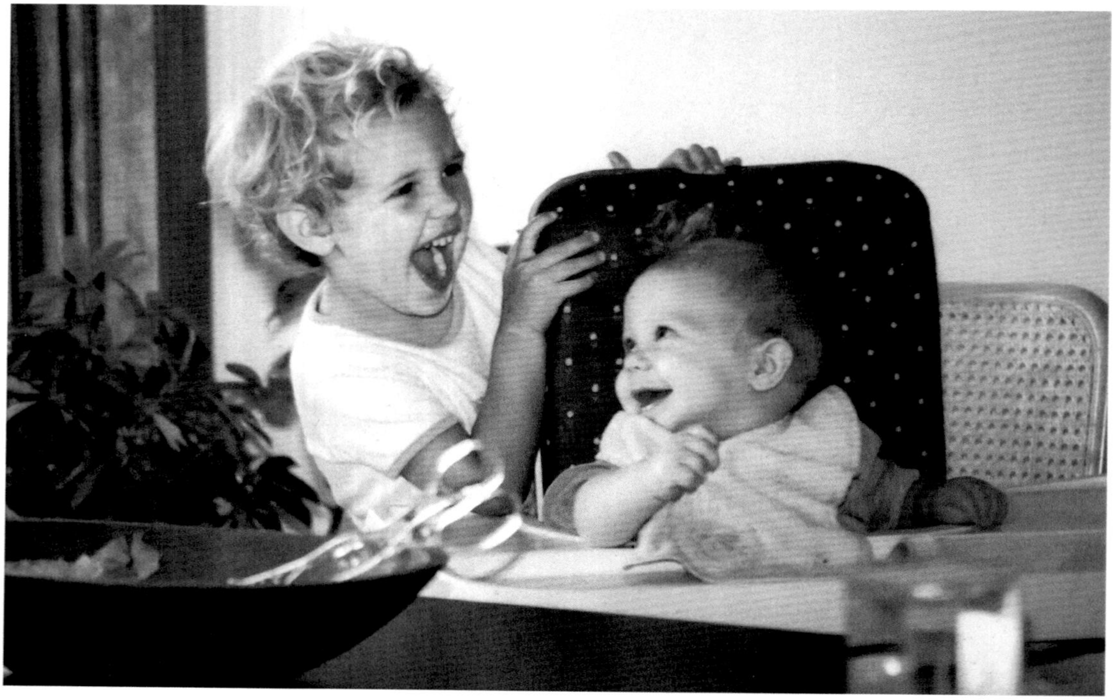

Figure 5-33. *Sisters. The implied diagonal makes the connection between them, while their expressions spell out their relationship*

Here the body language tells the story. The grins and the diagonal implied by the powerful eye contact between the sisters energize the picture and describe the relationship (at least at that moment; they're friends now, but they fought like snakes growing up.)

Photographers – especially wedding photographers – routinely try to enhance the story-telling quality of their work by photographing couples holding hands, couples stuffing cake in each others' mouths like Alvin on a sugar binge, couples in formal dress sitting by a poolside, splashing their feet while grinning delightedly at one another. Perhaps the couple is photographed silhouetted by the setting sun, or with her veil streaming out behind her in the evening breeze.

I'm not saying these traditional cliché shots don't work – they do work, or they wouldn't be used. And used. And used. The problem is that the photographers are trying to create a specific, romantic story that reads as predictably as The Little Mermaid! What they may NOT be doing is finding the real story as opposed to the clichéd construct they (and usually their clients) expect. Why not? Satisfying clients is what they're getting paid to do. It's not a crime. But relying on clichés limits the photographer's creativity and ability to create something really special for the client.

My point is that you can manipulate a scene to produce the story that you want, but the far more powerful stories are captured by paying attention to the small details and to what's going on in the scene.

Techniques

Good story-telling photos do not happen by accident (well, most of the time, anyhow). As a photographer there are techniques you can use to create a story-telling picture.

Composition. Great story-telling shots DO capture something immediate, in-the-moment, and authentic. They do it by paying attention to the traditional elements of lighting, composition, exposure, etc. Great story-telling pictures don't necessarily have to be great art – but there's no law saying story-telling shots can't be extremely artistic shots as well.

Here's an example of a story shot that is also artistic. Note the spiral composition, bringing us to the parachute jumper. The geometry is so strong in this shot that, coupled with the pattern of waves and color, makes the whole thing nearly an abstraction – but it's not, it's a story picture.

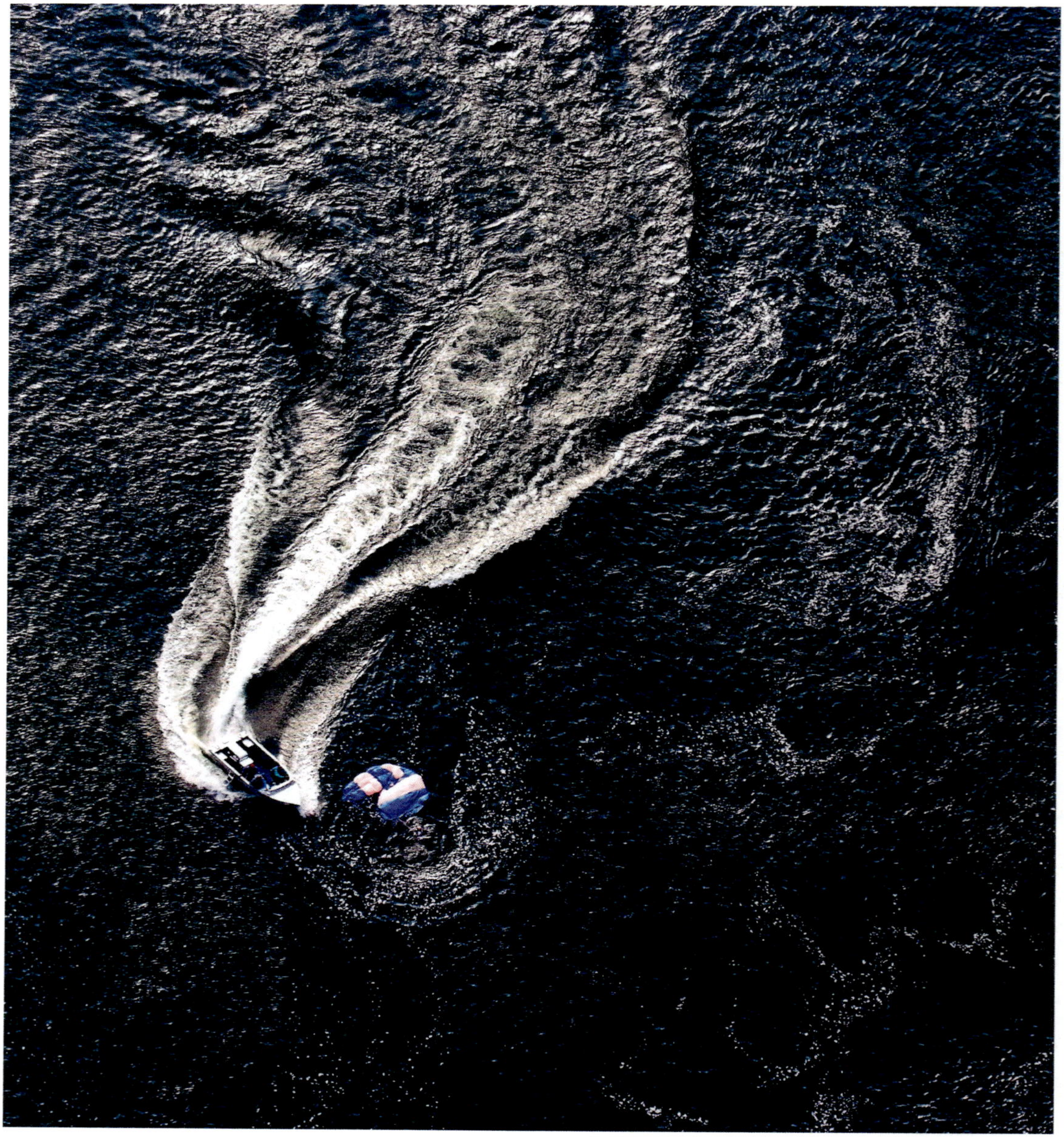

Figure 5-34. *Quick response from the water rescue team*

The story has characters, suspense, and action.

Compositional elements that create energy are diagonals, swirls, swooshes, S-curves (serpentines).

Rule of Thumb: anything off balance has more energy than anything static or at rest. There's a reason great sculpture rarely shows anyone standing flat-footed.

If you want energetic story-telling photos, composition does it. Bulls-eye composition, dead-center shots (with the exception of certain macros and portraits) tend to be boring. Good story-telling shots are never boring.

Focus on what matters. Partly this is a question of composition. While story-telling photos do need context, they don't necessarily need a lot of it. The "hands" pictures show examples of story-telling pictures that have enough information for context, but still have a central subject of interest.

Focusing on what matters is partly a matter of cropping, either in camera or afterwards, but it can also be achieved by two specific in-camera techniques: depth of field and motion blur.

Depth of Field. You can use depth of field to isolate your subject without eliminating the environment. Use apertures under f4.0 to get the most out of this technique. But don't over-do it. The check-signing picture above was shot at f2.8 from very close up; notice that the check goes out of focus at the critical points so no personal data is revealed. This was done in-camera, not in Photoshop.

Motion Blur. Motion blur creates the illusion of speed. Sometimes the speed is real, sometimes it's enhanced. Here are two shots, one caused by shutter speed, one by panning. Panning shots are achieved when you use a relatively slow exposure (say, 1/60 of a second) while keeping the moving body centered and steady by turning your body at the same rate as the subject. In the final picture, the subject appears to be stationary while the background is rocketing by.

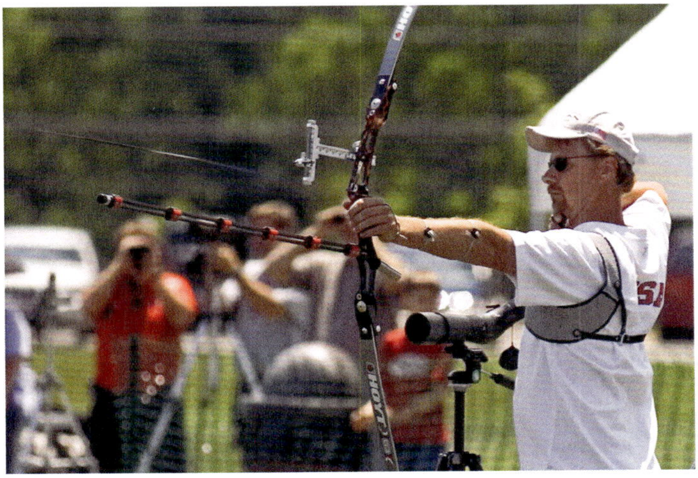

Figure 5-35. *Arrow in flight.*

This picture also uses depth-of field to help isolate the subject. It took over 200 tries to get this shot, because when champion archers release an arrow, they try not to move anything ... so you can't tell it's going until it's gone.

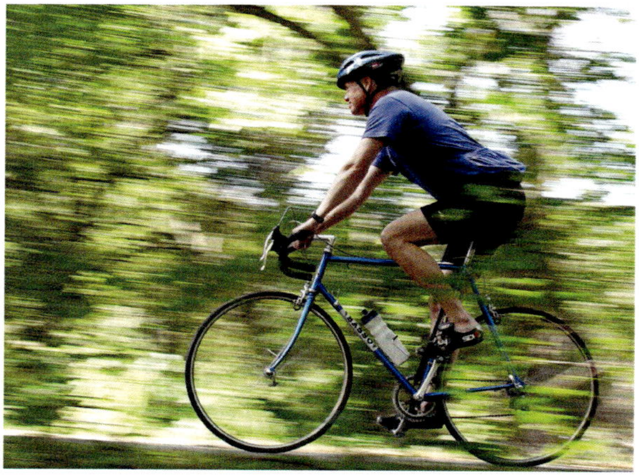

Figure 5-36 *Biker headed for home.*

Figure 5-36: The biker is really going a reasonable pace (as his posture shows), but the motion blur from panning emphasizes speed, while his expression shows how satisfied he is with the ride. The panning action causes the blur and serves to isolate the rider.

Lighting Control. This is the hardest thing to control, especially if you're shooting in lousy light at a sports event or wedding. But there are some things you can do to enhance your photo's story-telling strengths.

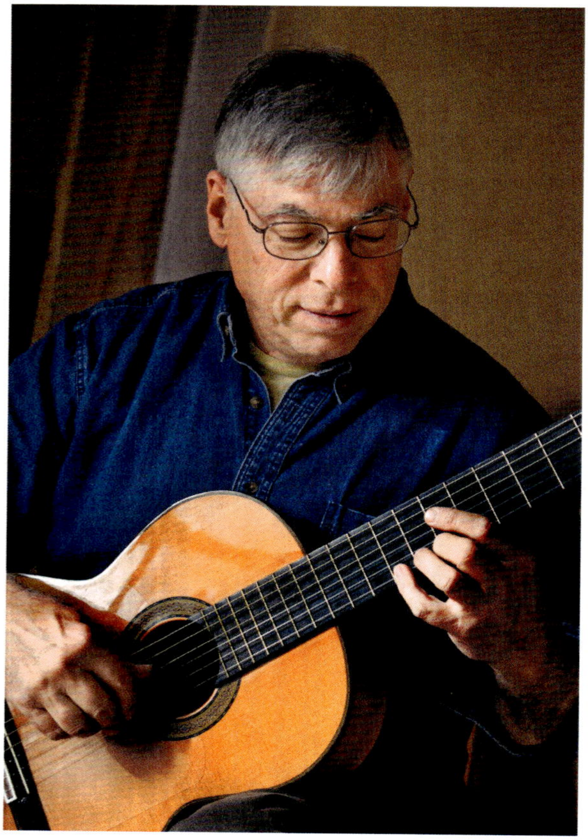

Figure 5-37. *Passion guitar*

Here are some tips:

Use natural light to emphasize and shadow to de-emphasize areas in your photo. The example here is arguably NOT a story picture, but it has many of the necessary elements. Shot in available light. Not how the fingers are emphasized, while the quiet lighting elsewhere creates a contemplative mood. The story here is the love affair of the man and his guitar. It's a relationship all right, and they'll never get divorced.

Use filters to intensify the photo and to put the emphasis where it belongs. Polarizing filters are really helpful to increase drama; graduated density filters can tame an over-bright sky. It's better to use them in the field than fudge them in Photoshop, but grad filters do work well as a Photoshop manipulation.

Wear a broad-brimmed hat. Best and most portable way to block unwanted light or sun glare when shooting out doors.

Use fill-flash to improve balance and focus on what matters. I typically set my fill flash to 1/3 to 1/2 stop under the non-flash exposure setting. The idea is to fill in, not eliminate shadows and detail.

Anticipate the Action. Sports shots absolutely require this. Volleyball, soccer, football, parachute jumping (yes, that again!) all work to one principle: if you wait for the shot you want to occur, it's too late. This is a simple matter of physiology. Humans with fast reflexes take about 1/10 of a second to see an action and send a message out to your finger to PRESS THE DAMN BUTTON NOW! So somewhere between 1/10 and 1/4 of a second have gone by, and in action-related work that's way too long! So you have to anticipate what's going to happen, and shoot at the right time. This takes lots and lots of practice, but it can be learned. The more you know about the particular subject you're shooting, and the more you practice shooting it, the better you will become. Expect a lot of failures and trash shots until you start getting it right. Electrons are cheap, so just toss out the crud and keep shooting.

Plan the shot. If you know what story you want your shot to tell, you can set things up so you get the results you want. While story shots are best if they're spontaneous, you may load the dice by posing your subjects, selecting your angle ahead of time, thinking about your location, and consider what kinds of lighting issues you're likely to face. Bring reflectors and light-blockers (gobos) if you need them. Know what lenses you'll need, and have them loaded. Being short, I like to bring a sturdy wooden step-stool to get up above the surrounding heads in a crowded location.

Summary

Story-telling is an element in most great shots. The borderline between a story-telling shot and a non-story-telling shot can be obscure, but in nearly all cases if you think about what story your image might invite viewers to tell themselves about the shot, you will add depth, energy, and emotional impact to your work.

PART TWO

Practical Advice

CHAPTER 6: *Preparing for a Shoot*

CHAPTER 6: *Preparing for a Shoot*

MATCH EQUIPMENT TO THE TASK

Cameras

I can't tell you how often I am asked, "what's the best digital camera for me to buy?" I can't possibly answer that, so what I do is sit down with the person and ask them a slew of questions that enable them to proceed. Here's a sample from a couple of weeks ago.

Amy works at a private school near Cincinnati.

Amy: I'm thinking of taking up photography. What kind of camera do you recommend?
Eric: That's great! What kinds of things do you want to photograph?
Amy: Uh, I guess mostly family, when I have one, and the kids at school.
Eric: OK, mainly people pictures, right? Are you interested in little kids or older ones?
Amy: Uh, both, really.
Eric: Where will you do most of your shooting, indoors or out, or both?
Amy: Mainly indoors. But some out, too.
Eric: How about the sports teams?
Amy: Oh, definitely!
Eric: Do you want to make prints, or mainly web use?
Amy: Mainly web, Facebook, like that.
Eric: Anything bigger than 8x10?
Amy: I don't think so.

At this point, Amy has told me more than she knows. I can now help her list the camera features that will matter to her.

- Resolution is not a big deal, 6-8 megapixels is plenty (she won't be making big prints).
- Hot shoe and external flash are really important if she's doing lots of indoor pictures of kids. Built-in flashes won't cut it.
- Shutter lag is an important consideration: little kids and sports teams both move in the blink of an eye, and by the time a slow camera shoots, it's too late.

- Fast autofocus capability matters for the same reason. This, I tell her, means more money, but will make a real difference.
- Good zoom lens for sports and kids running around outdoors, or on the stage indoors.
- A viewfinder, not just a display screen, is essential.

I explain some basics so Amy understands (dimly, at least) what sorts of features are critical for her intended use. Then I tell Amy she should consider a very good point and shoot, or an entry-level DSLR. I suggest she visit the http://www.dpreview.com site and use the camera reviews (exhaustive and kept current) to narrow her choices, then hang out in the Forums there to see what problems people have had with them.

I also suggest she join a local camera club like the Ohio Valley Camera Club and ask for help and suggestions.

Amy is appreciative, but far from satisfied because she's realized some uncomfortable truths.

1. She has to learn a lot, much of which is strange to her
2. No one will make her choices for her
3. She can spent a lot of money and not get what she needs, or spend a reasonable amount and be more satisfied – or the other way round, unfortunately for her.
4. She'll need to spend time and effort to find her camera
5. Once she has it, she'll need to learn to use it. But that is what she swears she wants to do, so OK.

The point of the exercise, and of this article, is that you have to match your equipment to your needs. There's no point at all in throwing money away on stuff that is far more or far less capable than you need.

The need to ask the same questions I asked Amy applies, however advanced you are:

- What will I be shooting?
- Where will I be shooting?
- What kinds of lighting requirements will I face?
- What will I do with the images once I have them?
- How important are prints, and how big will I need them to be?
- Will I need to beam my shots wirelessly (news and sports shooters may well need this capability)
- What is my budget?
- Do I need remote control?
- Do I need to shoot while tethered to a computer? (Product shooters take note)
- Am I interested in architectural photography? (Raises questions about lenses and resolution)
- Am I a birder? (Huge lenses, specialized flash equipment, excellent tripods all go with this particular obsession)
- What lenses will I need, and what will they cost?
- How will this equipment pay for itself? When will it have paid for itself? Does it need to?

- What will it allow me to do that I couldn't do without it, and how much of that do I realistically expect to do?
- How soon will it become obsolete?

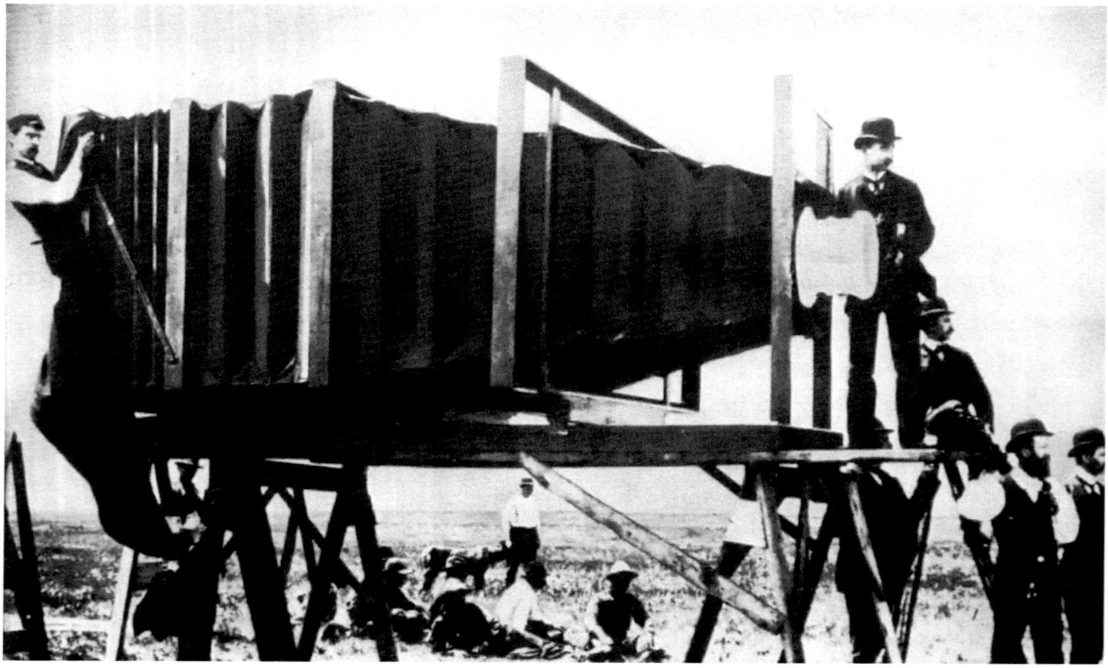

Figure 6-1. *Mammoth Camera. Do not take this baby to a wedding, sky diving, or parachute jump. The camera needs to suit the job.*

The more of these questions you ask before you even think of looking at gear, the more satisfied you will be with your eventual purchase. Of course, we all know that gear-lust overtakes us all at some time or another. Maybe I'll make enough from this book to get the Phase One IQ180 with the 80 megapixel back I really, really want!

Reason only takes us so far.

That said, I'm going to work through some of the basic considerations in buying camera bags and tripods.

Camera Bags

Camera bags are like the Middle East: warring factions everywhere, and compromises are hard to find. Basically, there are backpacks, shoulder-hung bags, belt-mounted bags, pistol or chest-mounted holsters, hard cases, and rolling cases. Each has advantages and disadvantages.

Questions for camera bags are the usual: what do I need to tote, over what terrain, for what distances. How fast do I need to get at the camera? How many lenses do I need to carry? And so forth.

The point is that you approach bags the same way you approach any gear question: what do I need it for? What will I do with it? How often do I need it? What can I afford?

Backpacks come in a wide variety of sizes. They can hold quite a lot of gear and pad it well, and they allow the weight to ride high on your back while your hands are free. But you have to take the pack off to get at your camera, and you need to find ways to attach your tripod to the bag. Some make this easy, some don't Another problem is handling a backpack for camera gear AND a back-pack for overnight camping.

Shoulder bags carry less than backpacks (if they are big enough to hold plenty of gear, shoulder bags a real pain to hike with because they pull your weight off-center, fatigue your arms and shoulders, and can cut into your neck). On the plus side, they're convenient for day hikes when you don't need a slew of different lenses, and they allow you to get at your camera more easily.

Belt mounted bags require, gee, guess what – belt loops, or an independent belt like a utility belt for cops. They're best for light weights, but they do leave your hands free and permit easy access to your camera.

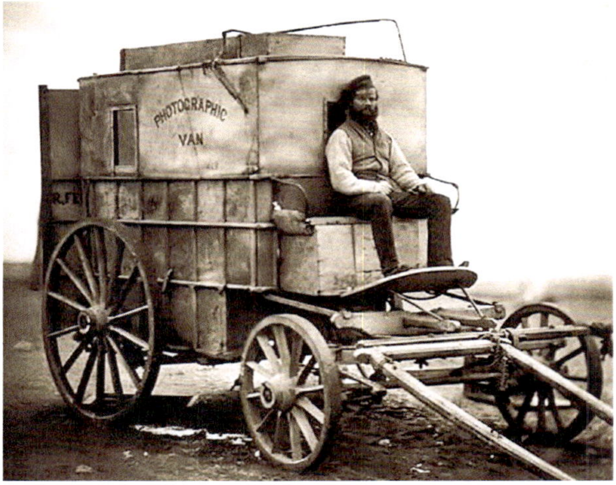

Figure 6-2. *A prototype rolling photo case, circa 1863*

Hard cases protect your gear. Pelican cases, especially, are the top of the line for gear protection. You can toss you most delicate camera, encased in a Pelican case, over Niagara Falls and expect to pop the case open and start shooting – if you can find it again. But Pelican cases are a pain to tote – I know, I've done it, down a cliff to the Pacific Ocean in Washington State. Add a tripod, and you're in trouble.

Rolling cases are great if you're doing urban work and don't have to deal with mud, water, sand, camel turds, or other impediments to easy travel. Even cobblestones make rolling cases unpleasant. But they do take the weight off you and protect your gear.

I haven't personally used either chest-mounted rotating holsters or belt-mounted holsters. If you're not changing lenses and need quick, stabilizing access, they might work for you.

So we're back to square one: what you choose depends on what you need. I like to shoot with two cameras, a Nikon D800 for fast stuff and a Mamiya 645 medium format with a 22 mp digital back for landscapes. I've tried jamming them both into an oversize backpack, I've tried a backpack plus a shoulder harness, and I've tried a Pelican case plus a backpack. No combination is really great – IF you want to be prepared for a wide range of shooting conditions. Otherwise, it's best to pick one camera and a couple of lenses and pack accordingly – knowing you'll miss out on some shots you'll wish you'd taken.

My real advice is TRY IT OUT before you buy. Rent a case, or just put 12 pounds or so of rocks in one (in plastic bags so they don't scratch) and go for a demo hike.

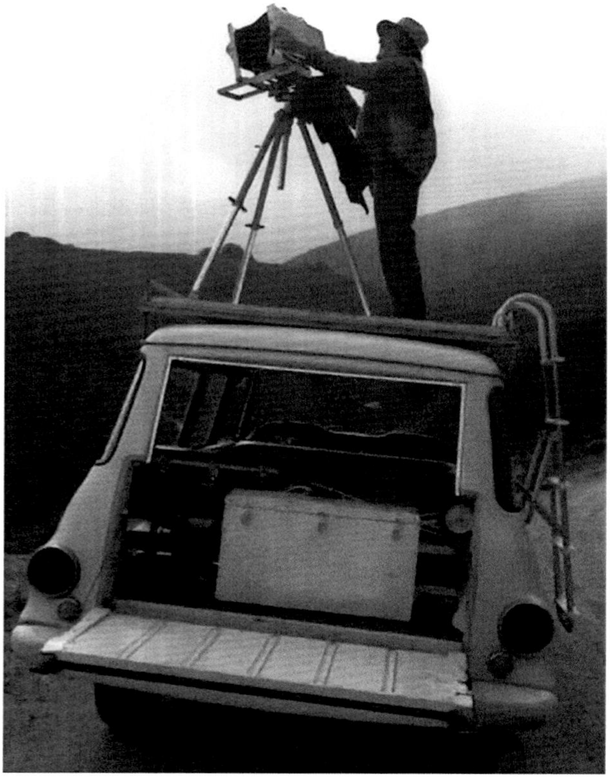

Figure 6-3. *Ansel Adams knew the value of a sturdy tripod.*

Tripods

Here are the key questions for tripods.

- What kind of shooting do I need it for?
- How far and over what terrain do I expect to carry it?
- How heavy a weight must it support? (1 pound for point-and-shoots, 2 pounds for DSLRs, 3-5 pounds for DSLR's with ginormous lenses, 5 pounds for medium format gear).
- How important is the weight of the tripod (including head)?
- Do you need a built-in compass and spirit level? (yes, if you're doing panoramas).
- What sort of head do I want? Ball head, pistol-grip, or 3-lever axis adjustment?

- How big a ball head do I need (bigger is better, but more expensive and less convenient). If you're using 300mm f2.8 lenses, the answer is "huge." What happens is leverage of the lens makes the head slip, or makes the camera slip on the head. Both are bad.
- How important is vibration resistance?
- How sturdy is it?
- What does it weigh?

I have a Gitzo carbon fiber tripod which is about 3 pounds lighter than the very sturdy aluminum tripod I used to use. That's a huge difference. But the feet on my Gitzo are not locked in, so they have a tendency to work loose and vanish. This louses up stability and causes vibration problems. So watch out for small details when buying a tripod!

Summary

All gear purchases should start at the same place, by asking yourself how you'll use it, what are the specific features you need for your kind of shooting, and how it will pay for itself, either in dollars or in satisfaction. Asking these questions, and living with the answers, is a hard habit to form, but it will add immensely to your satisfaction and save you a ton of money in the long run.

SPECIAL CONSIDERATIONS FOR SHOOTING LANDSCAPES

Everybody loves to shoot landscapes. And most of the time, everybody is disappointed in the results. This article addresses some of the problems in shooting landscapes and suggests some techniques to help. At the risk of repeating information which appears in a number of other articles, this article will cover finding locations, characteristics of good landscape shots, choice of lenses, use of tripods, and exposure issues.

I'll confess. I am guilty of the truly awful landscape picture below. I don't usually keep stuff this bad, but here it is:

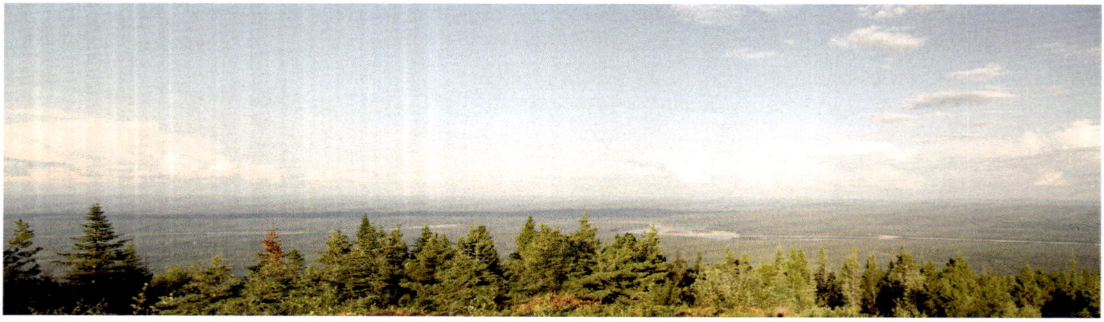

Figure 6-4. *Liard River, Yukon Territory*

Figure 6-3 is regrettably typical of landscapes taken every day by the well-meaning. It has everything wrong with it. Here's the list of atrocities:

1. Foreground very prominent, but the tops of trees are NOT interesting. There is no other subject of interest.
2. Background way too hazy to have significant detail
3. Far too much sky, and it's not even an interesting sky.
4. The Liard River, supposedly the subject, is basically invisible
5. Straight line composition with no variation to make it interesting – just layers stacked atop one another.
6. Color adds no value to the photo and in fact makes bad composition worse

Here are two other landscapes featuring "plenty of nothing," but oh, how different they are in character and quality from the embarrassment of Figure 6-4. Figure 6-5 depicts a truck traveling through North Central Montana. This image conveys the sense of vastness, but in this case there's a speeding pick-em-up truck, with a nice leading line and a serpentine dust cloud adding to the sense of action and wide-open space. Note that the horizon is well above the center of the picture; we're interested in what's happening on the ground. Moving the horizon line makes a huge difference in what the picture communicates! A graduated density filter would have put the sky into better balance with the rest of the image.

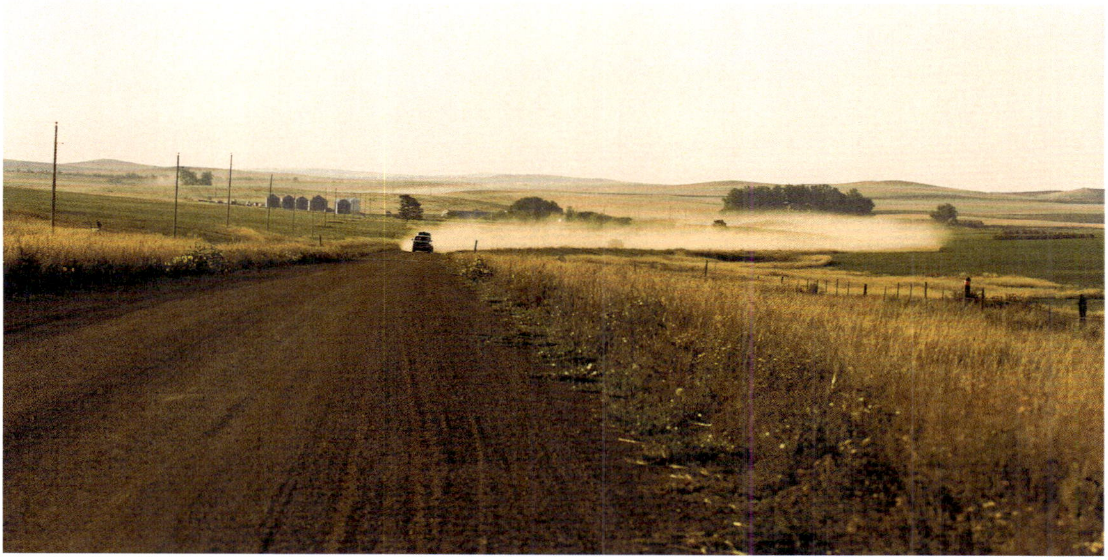

Figure 6-5. *Northern Montana dust cloud*

Action isn't required, but composition is. Here's another shot from the same part of the world (Actually, North Dakota):

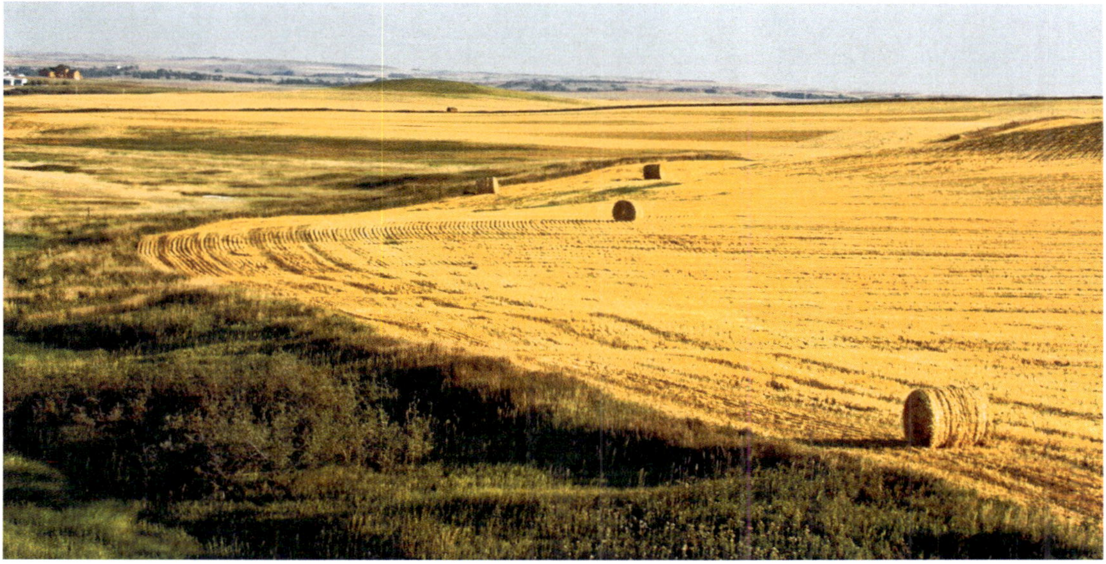

Figure 6-6. *North Dakota Hayfield*

In this case, contrasty late-afternoon light creates depth-producing shadows. The sensation of depth is enhanced by the S-shaped waterway snaking into the background, carrying the eye with it. The diminishing size of the hay bales also helps create perspective. In addition, the horizon line is set appropriately. The hay bales are the subject of interest, and they sit parallel to the green-to-gold break line so they work to enhance the sense of depth instead of fighting it.

Readers who have not yet visited "Composition Concepts" may look there for more on how composition makes all the difference.

So far we've seen one crummy landscape and two good ones. All three are in a panoramic format, but that works well for some landscapes and less well for others.

What do you want to achieve in a landscape photo? Above all, you want to achieve an honest sense of the place you're looking at. I'm not going to define honest – that's a very complicated matter when you're talking about photography—but I'm going let you wrestle with it on your own. But for now "honest" means being faithful to what's actually there and to the atmosphere or spirit you sense while you are there. (Hint: HDR is a good way not to be honest.)

To achieve that impression of honesty you need the things every photo needs: a subject of interest, lighting that enhances the story, effective use of color, the right perspective, well-organized composition, and a sense of depth.

One of the main differences between the first photo and the better ones is a huge difference in the sense of depth. Landscape photos, as you have gathered from my comments above, rely heavily on "tricks" of lighting, exposure, and composition to achieve the sense of depth.

Let's start with subject-of-interest and selecting your scene. You have to decide at the outset whether you want to try and capture the whole scene, or just a significant part of it. In literature there's a term called "metonymy," a Greek term meaning letting a part suggest the whole: factory hands, for example, where hands stands for workers. In landscape photography a similar concept is shooting ONE mountainside instead of a range of mountains, letting one stand for all, or a portion of a waterfall instead of the whole thing. This allows you to capture much more detail at normal print sizes than trying to capture a distant mountain range all at once.

You can, of course, go for the whole thing or take piecemeal bites. The work-around is to use a telephoto lens to let you capture more detail in distant objects, take a series of overlapping pictures, and stitch them to create a detail-rich panorama, instead of a line of bumps pimpling the horizon.

Here is a stitched panorama (5 exposures) that looks great printed on canvas:

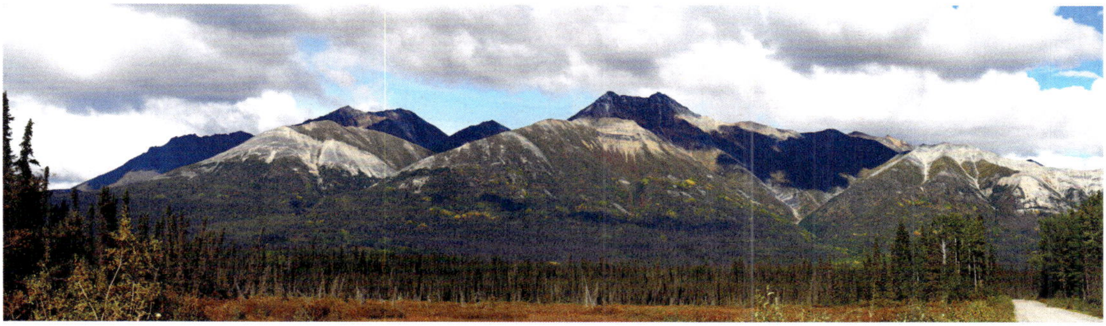

Figure 6-7. *McCarthy Rd, AK*

Note that the road is included to give a sense of scale. And, by the way, the sky is a little intense but the other colors are about they way they actually appeared (McCarthy Road, Alaska) But this article isn't about panos, it's about landscapes in general.

Here are a couple of samples of photographic metonymy.

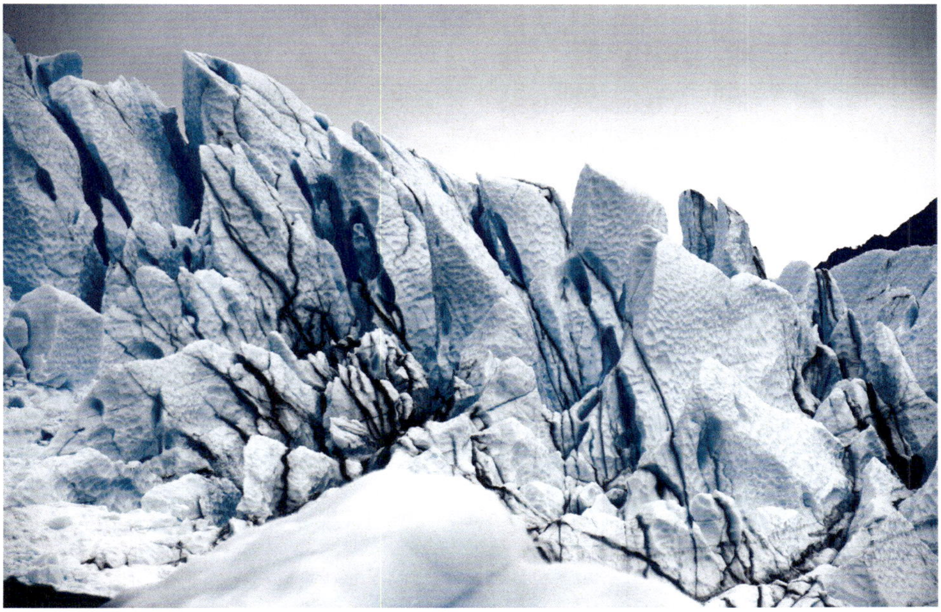

Figure 6-8. *Matanuska Glacier, AK - A closeup of a section of the glacier's leading edge*

And here's one that captures the whole scene:

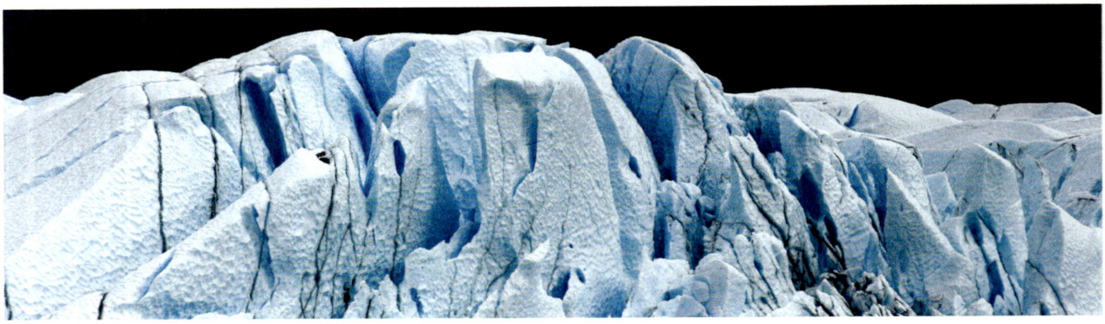

Figure 6-9. *Matanuska Glacier, AK - Showing more of the face*

The things to look for are a central subject of interest, regular patterns with gaps or intrusions to break the pattern – the eye goes right to such places in a photo and in life. The gaps and black lines (which are actually gravel) in the glacier photos are such irregularities, while the cracks are the more-or-less regular pattern. Both 6-8 and 6-9 are marred by flat light.

Try to avoid flat lighting. In the case of glaciers, clouds allow the blue tint to stand out, but normally strong side-lighting works best (the half hour before and after dawn and before and after sunset are best). Here's one shot the next morning before the clouds rolled in again:

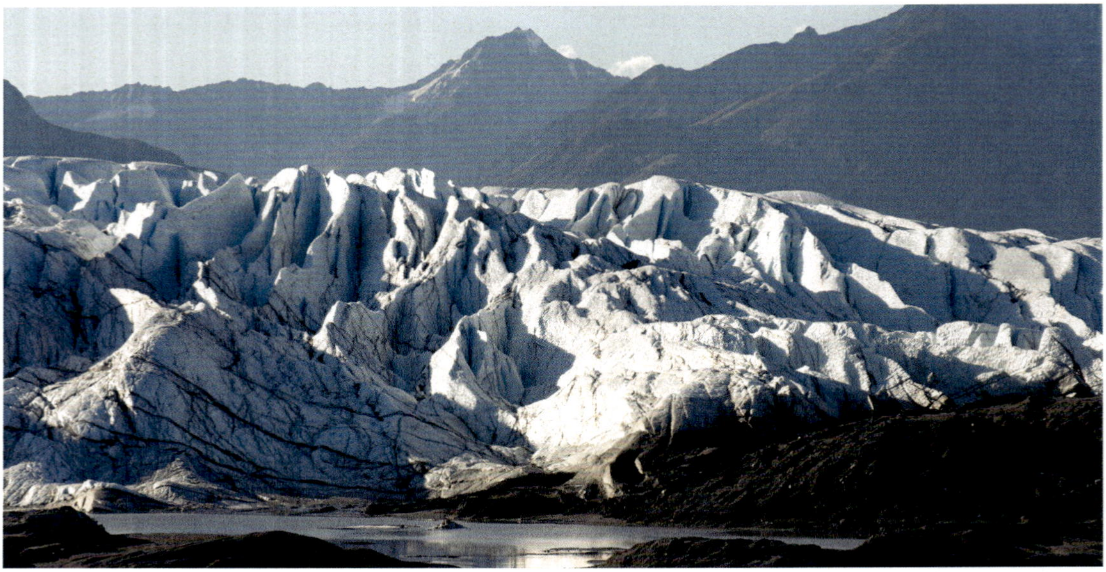

Figure 6-10. Matanuska Glacier in early morning sunlight.

Of course, you can only shoot big landscapes in the light God gives you, but God seems to favor those who are up before the sun gets too high in the sky. The shadows really help this photo, and the jagged mountain on the horizon makes your eye jump back and forth between glacier face and horizon, increasing the sense of depth.

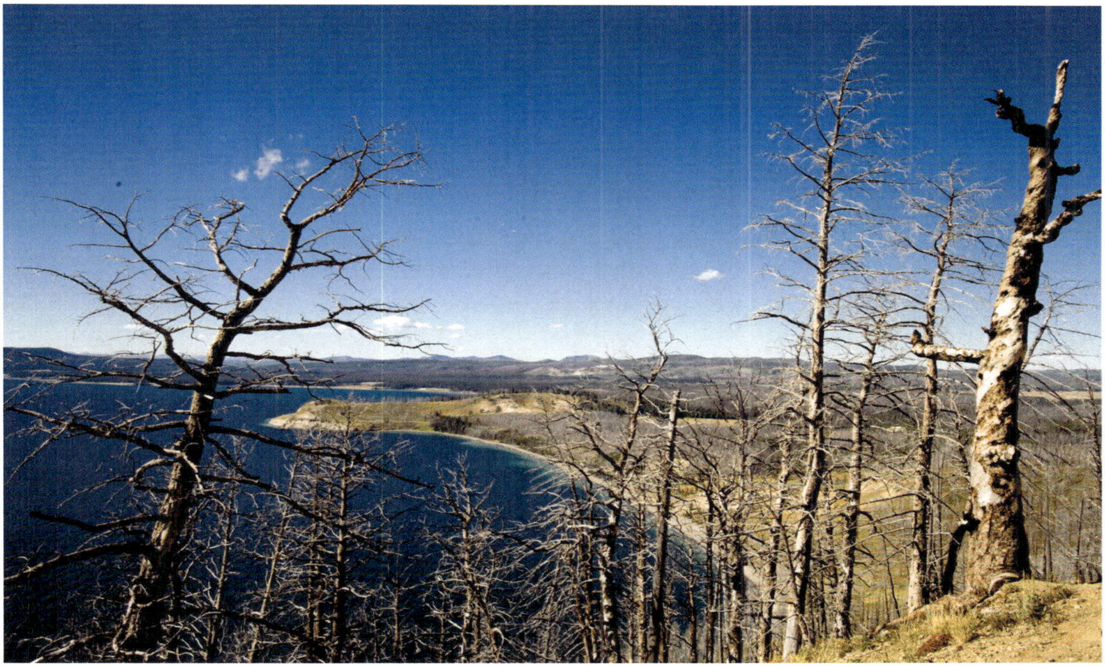

Figure 6-11. *Yellowstone Lake*

Trees are another opportunity to let the part stand for the whole. A zillion trees in one place look like a jagged green band, or brown fuzz, depending on the season and location. But one or a few standing out from the forest give the eye somewhere to land in all those fractals. That image at the start of this article shows trees done wrong. Here's another example using trees more constructively.

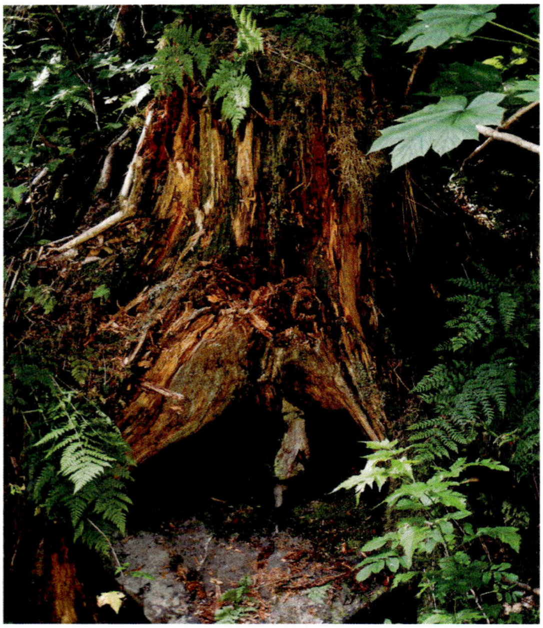

Figure 6-12. *Tree Stump near Seward, AK*

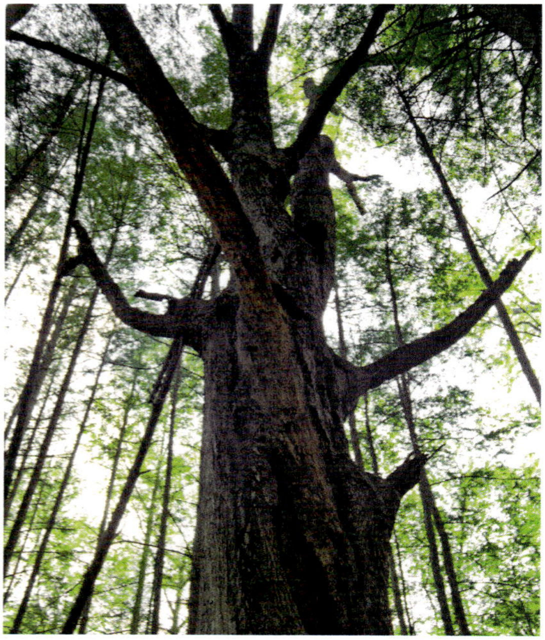

Figure 6-13. *New River, West Virginia*

In this image the two foreground trees set a frame and a mood, while the diminishing trees in the middle ground create the needed depth.

Here's an extreme example of metonymy. Would anybody have a problem accepting this picture if it were titled "Alaska Rain Forest"? Truly a case of letting the part stand for the whole in landscape photography.

It is time to move on to matters of perspective. I'm not talking here about the sense of depth, but of point-of-view. This is a matter of which lens you choose and where you position your camera.

Speaking generally, wide-angle lenses (under 50 mm) exaggerate distance by making objects in the immediate foreground appear enlarged, while reducing the size of distant objects. Telephoto lenses do the reverse; they foreshorten distance by bring far objects close, so that background and foreground appear flattened together.

With wide-angle lenses in particular, if they are not plumb vertically they will induce curvature and key stoning. In other words, if you shoot a wide-angle lens up into a grove of trees, all the tips will appear to bend towards the center. This can be a fun effect, but if you're shooting architectural exteriors, you don't want it! Here's an example, shot with a 25 mm lens. It's not a very good photo, but shows you what I mean.

Even with a normal lens, around 50mm, shooting with the camera tilted up emphasizes sky and lowers your horizon line; aiming it slightly down brings the ground into play and raises the horizon. Both are OK, depending on what you want to do.

Here's an extreme low-level shot (I was kneeling in the water) with a 15 mm wide angle lens.

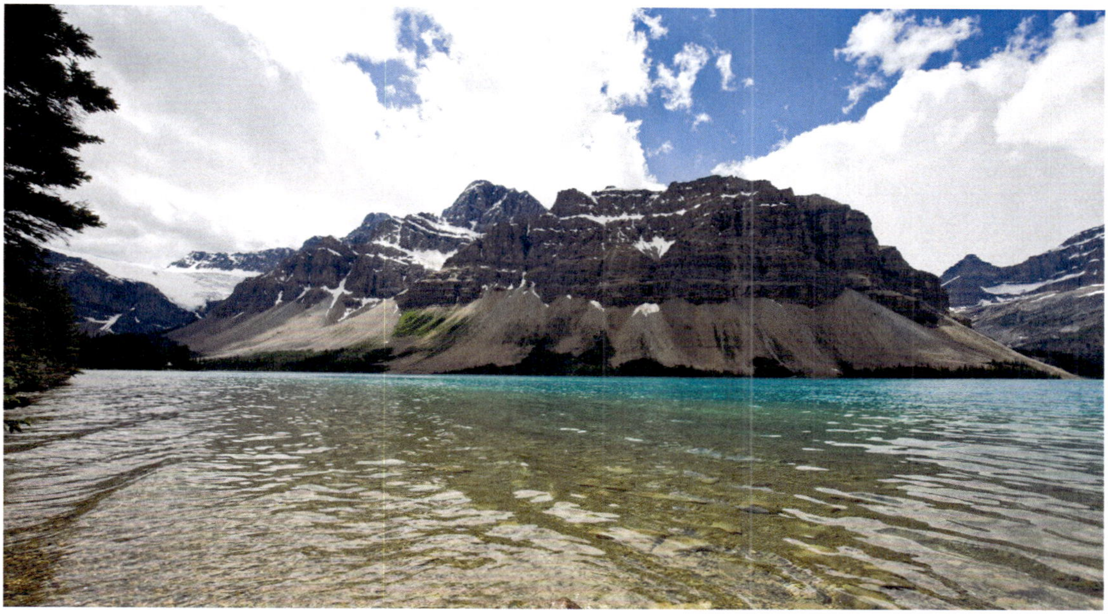

Figure 6-14. Glacier Lake, Icefields Parkway, British Columbia

The camera is held fairly vertical, but the exaggeration caused by the lens causes the clouds to act as leading lines, pointing at the mountain and its glaciers a little right of center; the opposite angle of the waves leads the eye left of center. This creates nervousness or tension in keeping with the restless water. It's a beautiful scene, but not tranquil or still.

You can use any of these lens characteristics to your benefit in landscape work. There's no room here to go through all the variations of tilt, roll, high and low. For now, just be aware that standing straight up and firing away at eye level may not give you the stunning shot you were after. Go experiment!

Vibration Reduction

When you are taking photos of distant objects, no single object gets a lot of pixels dedicated to it. Put another way, if you take a picture of a mountain with an interesting tree sticking out of it, the odds are that the tree will not take up a lot of room in your frame. Small in frame = few pixels devoted to it.

The consequence is that any sort of camera motion is a problem. You can get away with this for small prints, but in very large ones you'll see the effects of motion blur. That's because once you've enlarged the image, the innate fuzziness of motion blur is painfully obvious.

On bulb exposures (long exposures at night), the problem is magnified, and even the force of opening the shutter can induce enough vibration to ruin your evening.

Ways to reduce motion blur are

- Use a rock-solid tripod (which, of course, you have accurately leveled)
- Make sure the metal tips are not protruding from the feet of the tripod
- Use shutter pre-up if your camera permits
- Use shutter delay
- Use a remote electronic shutter control
- Avoid telephoto lenses for night shots. They magnify any camera shake at all.

PANORAMA PARTY (SHOOTING AND PREPARING PANORAMAS)

Over the years I've fooled around with panoramas. Lately I've been playing with them again, and you can see some results on my website. I have also been asking a few questions about panos and how they work.

Some panos are stitched images, while others are cropped single frames. Doesn't matter. What they all have in common is an aspect ratio of about 2:1 or greater. (Wide-screen TVs and monitors tend to be 16:9, which is a hair less than 2:1, but there's no need to quibble.)

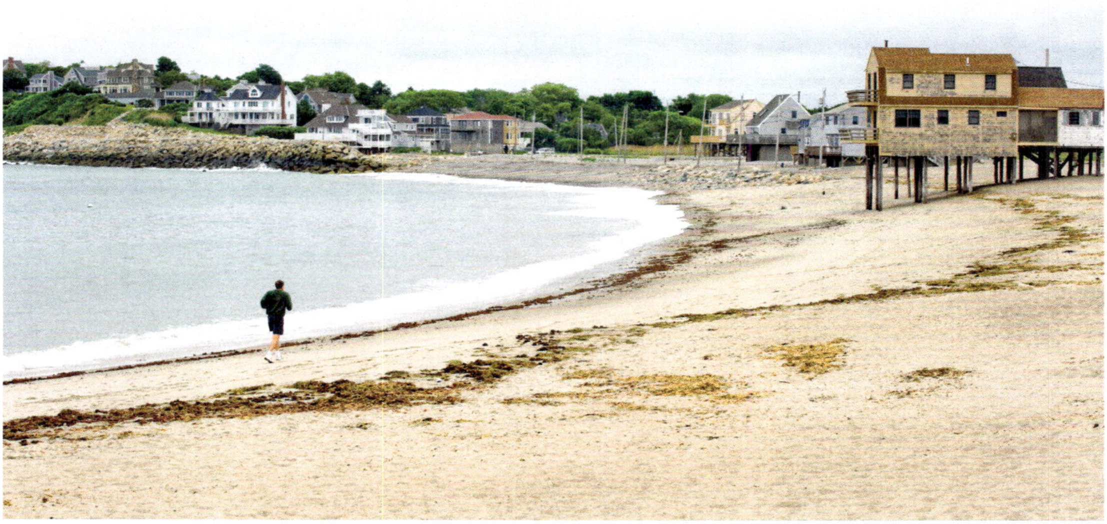

Figure 6-15. Runner near Scituate, MA

Panos seem to have a unique ability to grab our imagination, to move us through a scene. In this society, our eyes are trained to move from left to right, and from top to bottom. Panos greatly reduce the top-to-bottom component (Unless they're vertical panos), and our eyes travel the length of the image. Accordingly, panoramas can create a number of effects: they can create the illusion of time passing. They can emphasize the size or beauty of vistas, or mountains, or bridges. I haven't shot these subjects, but imagine a pano of a cheetah or a snake (yes, we've all seen the famous one). The pano takes the natural characteristic of an elongated object and makes us consider it a bit at a time.

Of course, not all panos are horizontal, and you can create dazzling ones from very tall elements -- but watch out for key stoning! Keep your distance, or use a perspective correcting lens (as in this Maui waterfall).

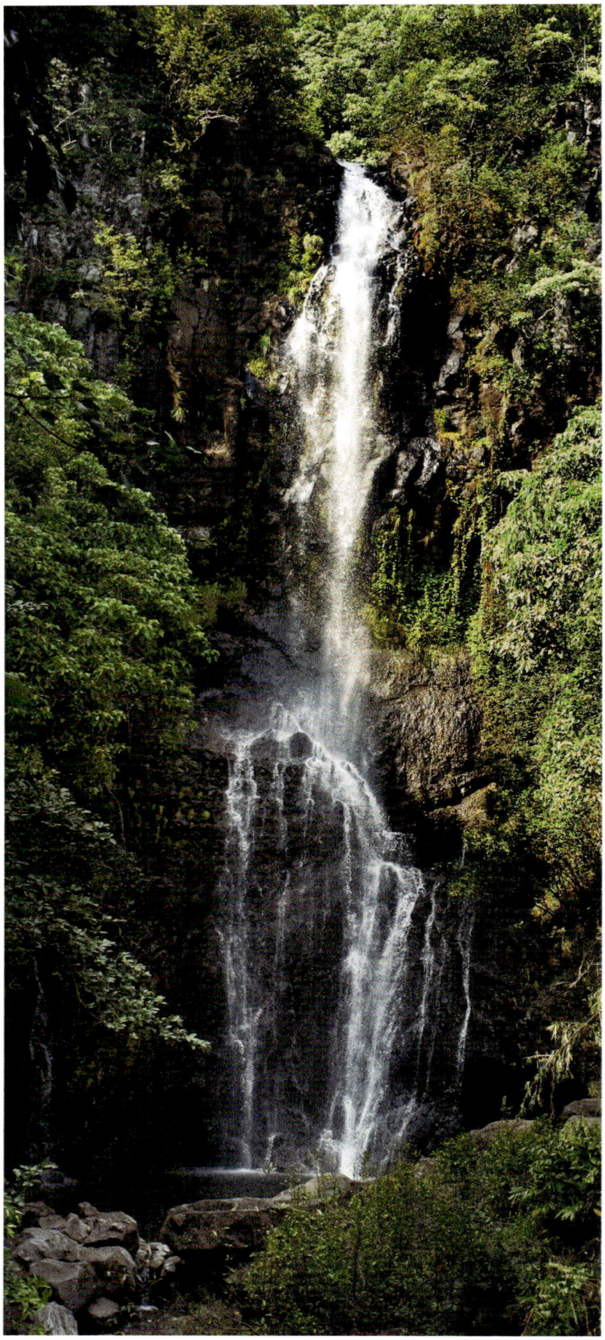

Figure 6-16. *Wailuku Falls, Maui, Hawaii*

Actually, shooting all your panoramic images in portrait orientation then stitching them into a horizontal landscape is a good way to get more "headroom" in any pano.

Some elements of a good pano are:

1. Single main subject (true of all photography)
2. Strongly linear nature of subject (vertical OR horizontal)
3. Plenty of headroom to allow for type (for commercial panos)
4. Spectacular color, exposure, and contrast to match the spectacular presentation of your subject.
5. Access to a great printer! Panos look good on screen, but they look best in large prints.

Basic techniques for creating panos:

Use a tripod. Make sure it is REALLY level. There are bubble levels built into many tripod heads for this purpose, as well as devices that mount on your camera. Special mounts are available which put the focal plane of the camera directly on top of the center of the center post. Expensive, but if you're really serious, they help.

At the risk of contradicting myself, it is possible to take good panoramas by rotating your whole body, as if you were taking a panning shot at a car race. But you'll get far better results with the tripod.

More tips for Shooting and Stitching panos:

1. Make sure your prospective pano has definite features that will help the software create appropriate overlaps.

2. Shoot in about 15 degree increments (a compass helps with this) to make sure you have enough software to do a good job. Many tripods come with markings for pano shooting, and some include a small compass.

3. Carry a small bubble level. This will really help you level your camera. The built-in ones on tripods are rarely exact enough to give you perfect alignment.

3. Shoot in manual mode to keep the camera from changing exposures from shot to shot. You don't want one frame that can make one frame noticeably brighter (or darker) than its neighbor, which often happens when the camera makes automatic exposure corrections.

4. If you don't use a tripod, try rotating your shoulders and not your neck from shot to shot. The idea is to put the center of rotation on the plane of the sensor, and this little trick helps you get closer.

5. You can try stitching uncorrected images, then correcting the resulting pano. This helps avoid problems with banding, notably in the sky. Or you can apply the same corrections in Camera Raw prior to stitching. I commonly expect to work on skies after the image has been stitched, regardless of which work flow I used.

Have fun, and try for the unusual!

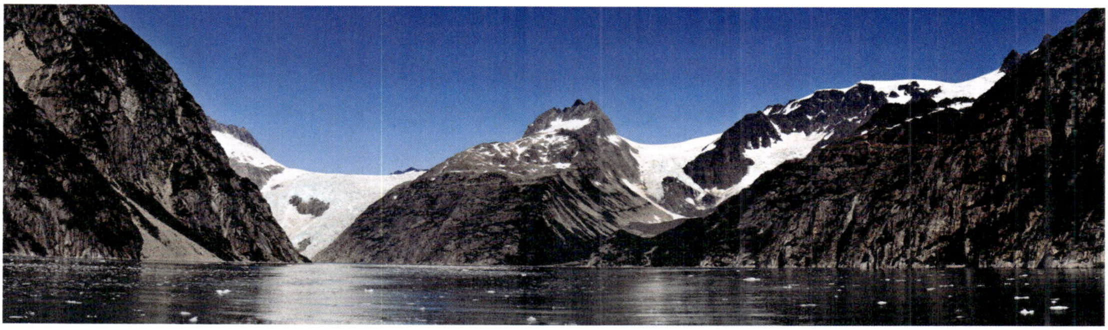

Figure 6-17. *Northwest Glacier, Near Seward, AK.*

CHAPTER 7: *Lighting in the Great Outdoors*

TOUGH LIGHTING SITUATIONS (BEACH, SNOW, DESERT, FOREST)
SHOOTING IN THE FOG
USING FILL FLASH OUTDOORS
SHOOTING IN THE DARK

CHAPTER 7: *Lighting in the Great Outdoors*

TOUGH LIGHTING SITUATIONS (BEACH, SNOW, DESERT, FOREST)

Shooting at the Beach

Ah, a day at the beach! Girls, gulls, waves, splashing kids, eager surfer wannabes – not to mention interesting debris (other than cans, 6-pack rings, and the occasional oil spill). There's a wealth of photo ops it's hard to resist. But beach shots can be a beach to shoot.

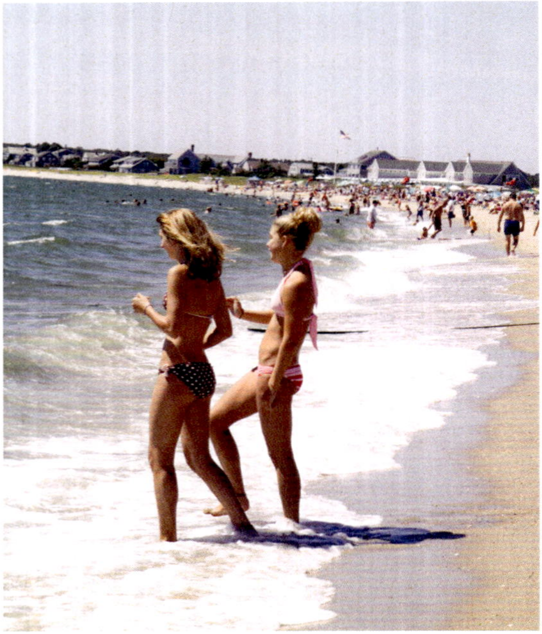

Figure 7-1. *Beach Scene, Chatham, MA No polarizer used.*

Generally speaking you'll encounter three lighting situations. First is the sunny day at the beach You'll have brilliant sand, bright sky, blue water. How do you get the shot?

This is a situation where old wisdom is still valid: Sunny 16. Shoot at f16, 1/250, ISO 100 or 160, and you should be in business. You can probably benefit from compensating up ½ stop or so. If you want precision, use spot metering off people, grey driftwood, or something else that's halfway between brightest and darkest in your frame. Her are four examples shot at about this setting. The lens was an 85 mm portrait lens (I used it because it has a lovely bokeh (smooth transition from in-focus to out-of-focus).

Both figures 7-1 and 7-2 are unpolarized, while 7-3 and 7-4 both used circular polarizing filters

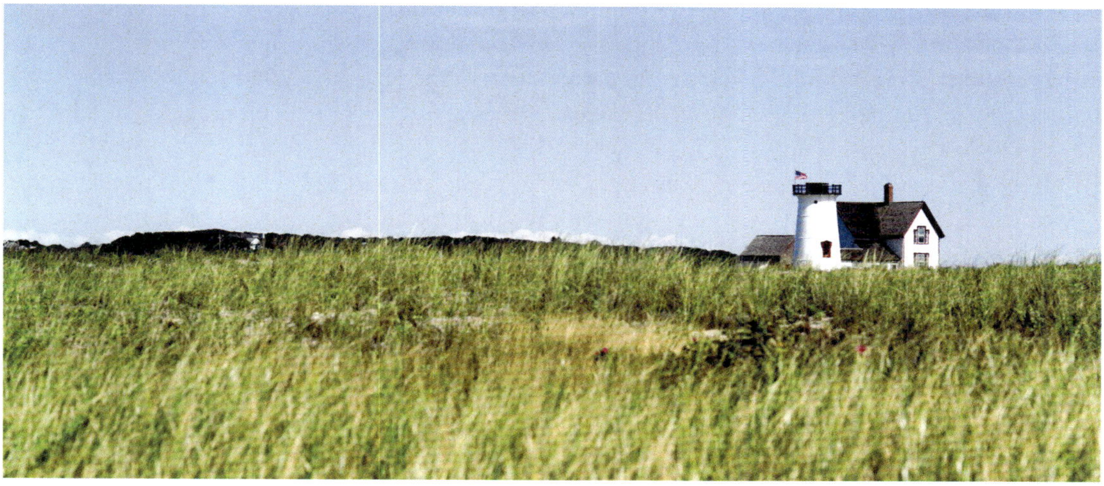

Figure 7-2. *Lighthouse, Chatham, MA . No polarizer used.*

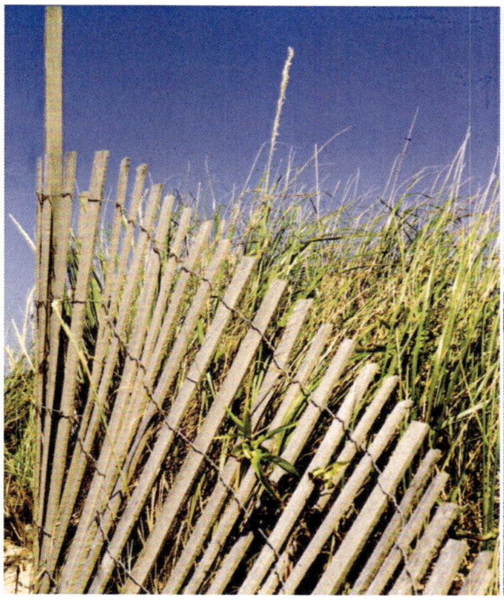

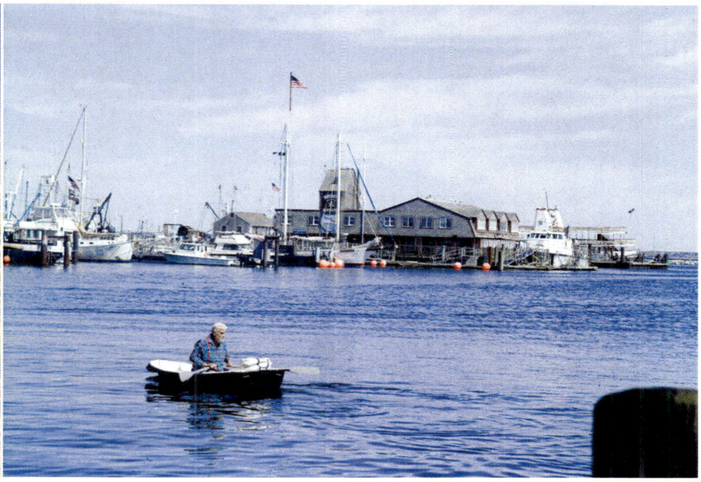

Figure 7-3. *Erosion fencing (Hyannis, MA)* ***Figure 7-4.*** *Man in Rowboat, Provincetown, both f16 with polarizer.*

There are three other common beach lighting situations.

1. Light overcast. This is wretched for shooting landscapes, fine for portraiture. The problem is too little contrast in the scene. Also, the sky is brighter than it appears to our eyes, so use matrix metering and plus up 2/3 to 1 stop (0.6 to 1 ev). The thickness of the overcast makes a big difference in exposure; you may need to experiment to find the right settings

2. Sunrise / sunset. These are wonderful conditions for beach shooting, unless you have bright sun just at sunset and dark shadows. If you're shooting people at sunset, you have some options.

a. If possible, let the light hit them at a 120 degree angle, or thereabouts (4:30 on a watch dial).
b. Let the light come in between 0 and 90 degrees, and use a fill flash to light up the faces (or other close-up detail in the shot.)
c. If the sun is still above the horizon, use a grad filter to bring the sky down, then plus up your overall exposure ½ to 1 stop. This will bring the people (or anything below the horizon line) up relative to the sky and sun.
d. Use spot metering — but you may sacrifice detail elsewhere in the picture.
e. Shoot in silhouette, and use fill flash again to brighten the foreground.
f. Play with sensitivity (exposure adjustment). I recommend using an external flash; your internal flash simply hasn't the horsepower unless you're right on top of your subjects.

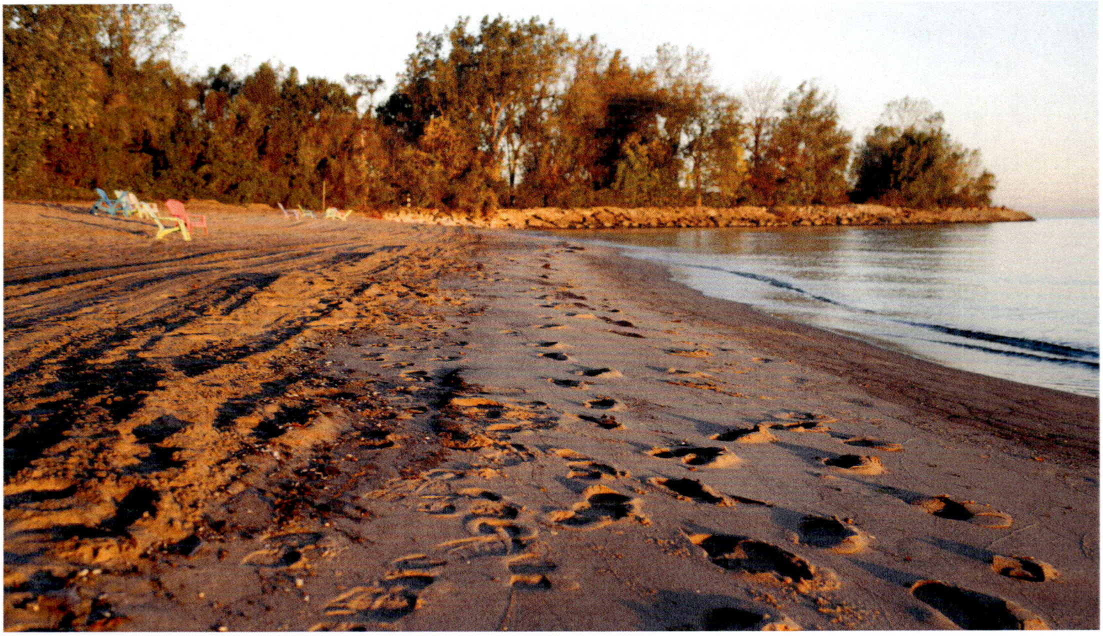

Figure 7-5. *Near Sandusky, OH. Lake Erie beach at dawn.*

Figure 7-6 was not well planned, but it works because there was sufficient light bouncing off the water to keep the man from turning into a silhouette. Maintaining shadow detail is particularly important in high-contrast shots. A grad filter would have turned this pretty-good shot into a superior one.

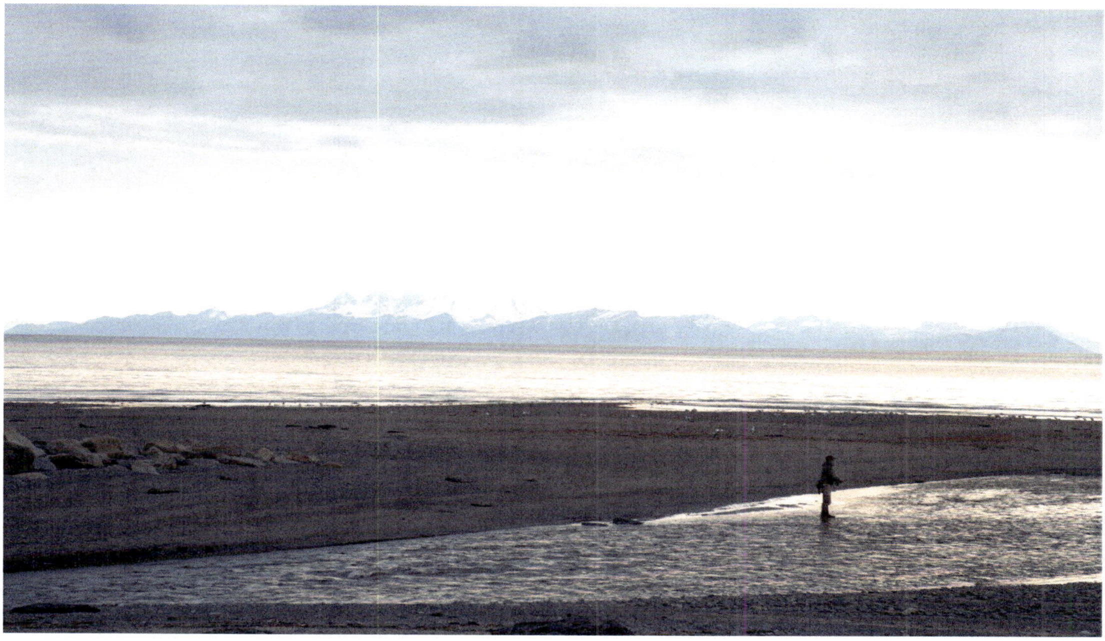

Figure 7-6. *Dusk fishing in Ninilchik AK*

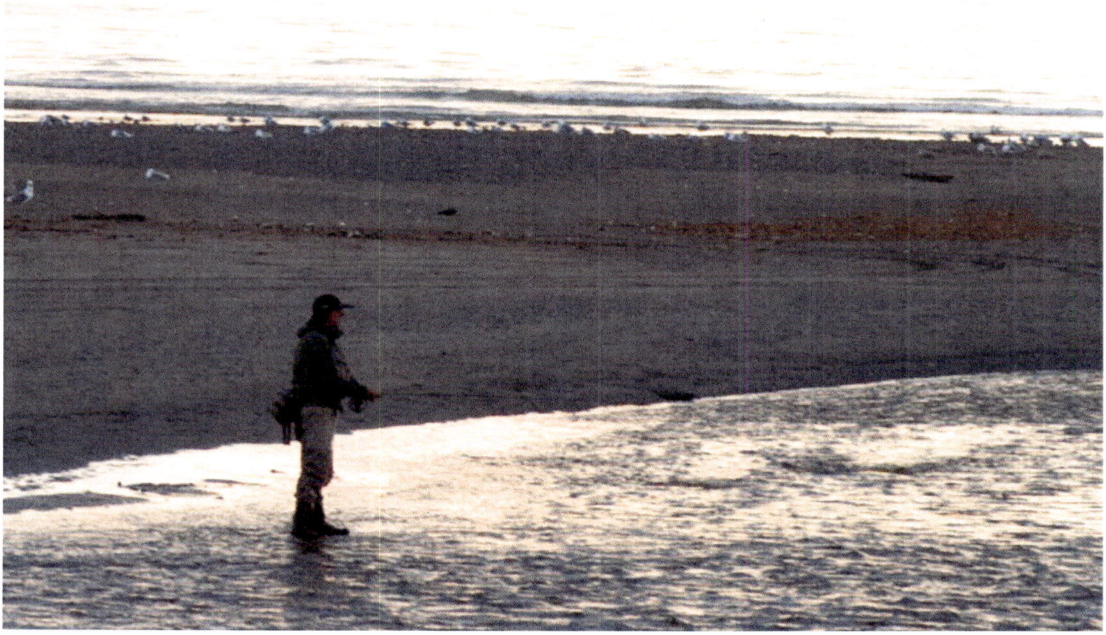

Figure 7-7. *There's just enough light to keep the fisherman from being a silhouette.*

Shooting in the Desert

The desert is the beach minus the water and, often, with more interesting terrain features. You have the same photographic challenges. There are really only two hours a day worth shooting in the desert: the half hour before and after sunset, and the half hour before and after sunrise. Early to bed and early to rise, makes a photographer wealthy and wise.

Shooting the desert in high sunlight is not only unrewarding photographically, in summertime it is plain dangerous. Never go into the desert without some form of protection from the sun and plenty of water. Leave word with someone where you'll be and when you expect to be back. Don't count on your cell phone; getting a signal is the exception, not the norm. I carry a pocket-sized SPOT satellite rescue transmitter for locations where there's real risk.

Unlike most beach shots, the desert can provide spectacular color; again, late or early sun makes the desert colors pop, as in these two shots taken on the same day about half an hour apart. These are straight-from-the-camera. (35mm camera on the left, medium format on the right)

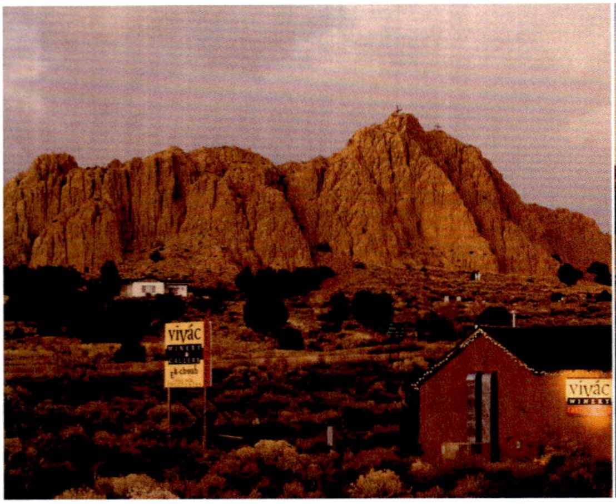

Figure 7-8. *After sunset, winery near Dixon, NM.*

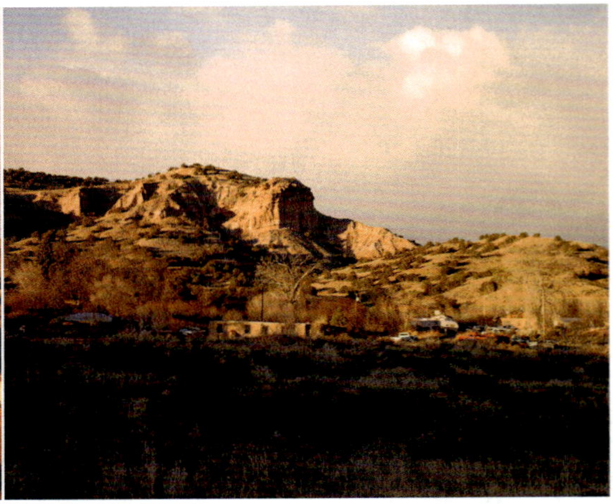

Figure 7-9. *Just before sunset, also Dixon, NM. Note how late light warms the color.*

One issue you face in desert shooting is a tendency to very black shadows. Also, unless you're shooting late or early, your camera is going to read the general scene and underexpose. It does this because it's trying to find a balanced shot in a generally very bright environment. So compensate upward by 1 stop to compensate, trot out your grad filter, but beware the polarizer. It will bring back some sky but it tends to intensify already dark shadows. So caution is needed. If it's getting really dark where you're shooting, reverse the compensation to minus .5 to 1 stop.

You have to decide which elements in a high contrast scene are essential and which are sacrificial. In general, I advise exposing for the mid-tones and shadows, especially when using either a grad or a polarizing filter.

Snow Scenes

Snow is not white. Because it reflects sky color and sun, it ranges in color from blue to gold to purple. And when it's not any of those, it's grey. Snow also picks up coloration from things that fall (or are poured) onto it.

These pictures show some of the different colors in snow. We have gold and blue in Figure 7-10, blue and purple in Figure 7-11, and orange and red in Figure 7-12 (Page 136).

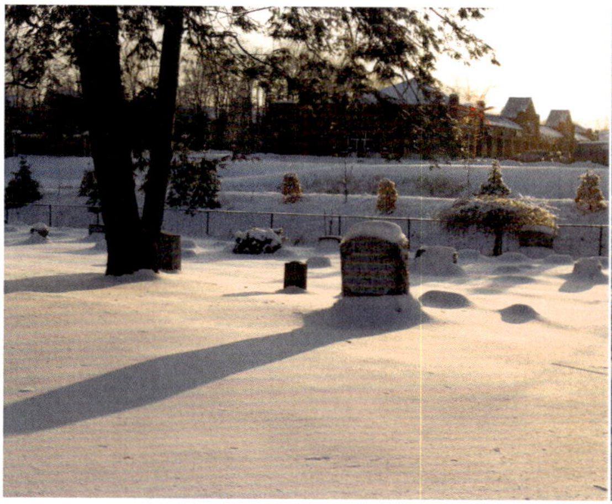

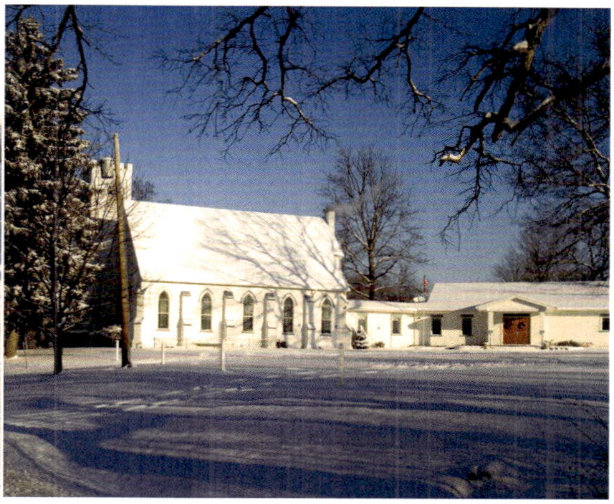

***Figure 7-10.** Graveyard, and **Figure 7-11.** Cozzadale Church. These images show the range of coloring snow acquires in different light against different sky colors. Snow is very rarely white!*

Despite all the color that snow can reflect, its still essentially monotone. That means you can get great shots in which a splash of color – a cardinal, a brilliantly colored parka – contrasts with the white. Lacking such a color contrast, consider converting to black-and-white to take advantage of the monotone nature of snow. It's a judgment call, like everything else in photography.

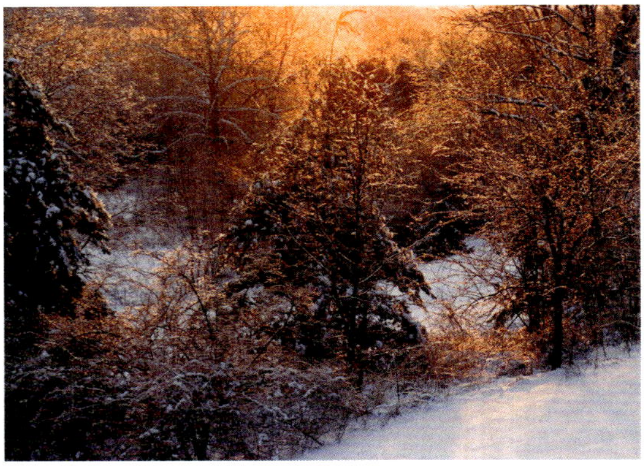

Figure 7-12. Ice storm sunset

Because of this color variability, when you shoot dictates a lot of what you get. Snow shots do not benefit from featureless stretches of blown-out white. It's important to expose so that you get at least a hint of texture in the snow. In the cemetery shot above you can see a sugary texture to the snow, and the faint play of shadow across the undulating ground enhances the sense of texture and depth.

Figure 7-13 does a good job of maintaining shadow detail inside the structure while keeping definite texture in the snow. It's one of my all-time favorite shots.

Two final points: if it's sunny, act (photographically, at least) as if you were at the beach. Also, despite the importance of maintaining texture in snow, your subject of interest has to come first. If it's your kid in her snowsuit, worry about getting her right and let the snow look after itself. Of course, if you're able to do both... go for it!

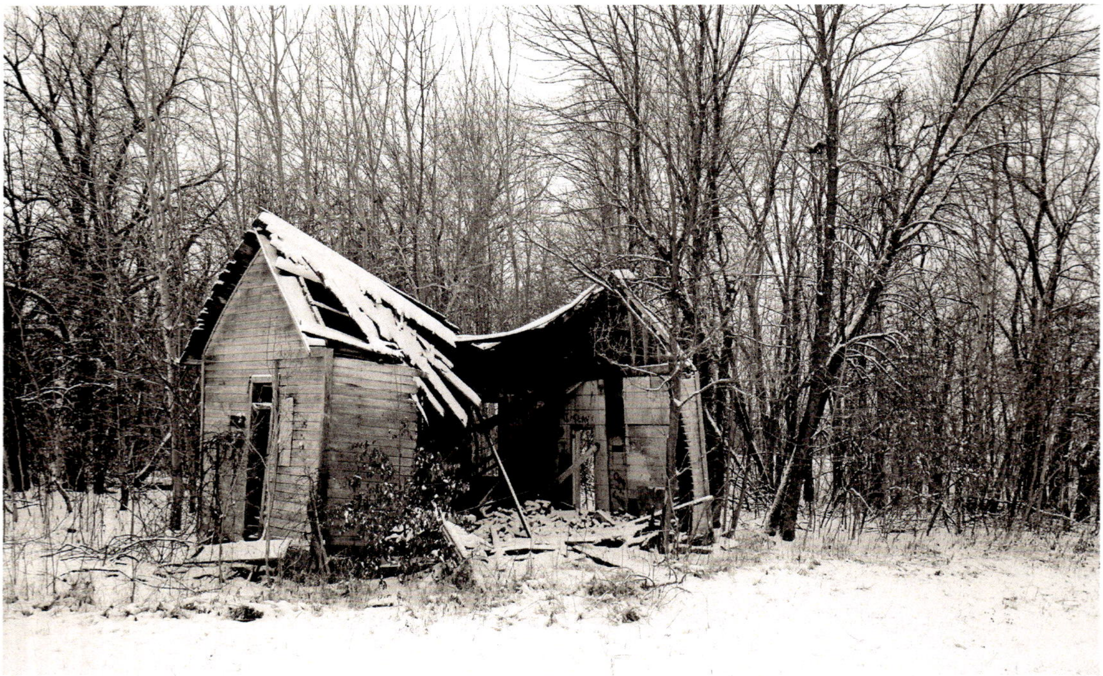

Figure 7-13. *Deserted barn in winter, central Minnesota. Detail is held in both the snow and the deep shadows.*

Finally, one last snow shot. Ice, compacted to the density of glaciers, is an intense blue. This is because the ice filters all other colors, and only the blue reflects to our eyes. This blue color shows up best, paradoxically, on cloudy days. But it can make for irresistible shots.

Figure 7-14 below shows all the texture you could want, because the light is raking across the ice. Note the serpentine composition, and the sweeping patches of light and cloud. Pray there is no downdraft while taking shots like this!

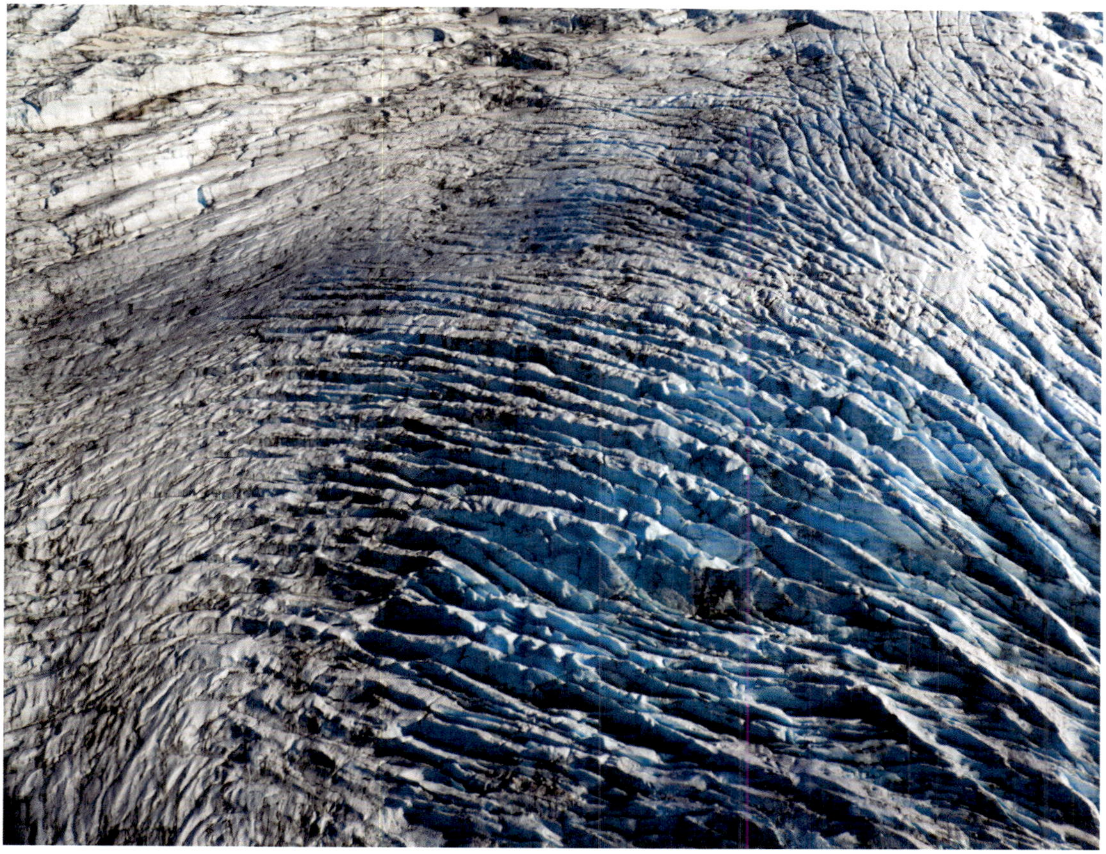

Figure 7-14. *Aerial of glacier near Kachemak Peninsula, AK. Side lighting enhances color and texture.*

Shooting in the Forest

Shooting in the forest is the reverse of shooting on the beach or desert. Here, shadows are everywhere and splashes of light make life difficult. The pervasive green color can be hard to capture accurately and creates a color cast that it's far easier to take care of in the camera than in Photoshop.

On the beach, you typically have to plus up your exposure because the camera will try to even things out and underexpose. In the forest it's the opposite: because the general environment is dark (particularly late in the day), you will have to correct your exposure by -½ to -1 stop. How much depends on how dark the scene really is; when there's no leaf cover or it's not too deep, you won't have to worry about dropping your exposure a bit.

What can you do about finding the right exposure? There are several options.

1. Start by shooting a grey card to set your color balance. Find a place in a clearing or by the trunk of a tree and set a custom white balance. If you don't know how to do this, see your manual (I know, I know...). As I explain in Chapter 9, color balance affects exposure, so the more accurate your color balance, the more accurate your exposure will be. Check the histogram. It should be basically a spike in the dead center of the graph.
2. Spot meter off something that appears half way between light and dark, and shoot in manual mode using that exposure. Gray human hair is terrific for this purpose! If you look at the histogram, do NOT expect to see any significant area on the right side of the graph. If there's nothing really bright in the scene, you shouldn't expect it to be an evenly distributed histogram.
3. Bracket your exposures.

Here are three forest shots.

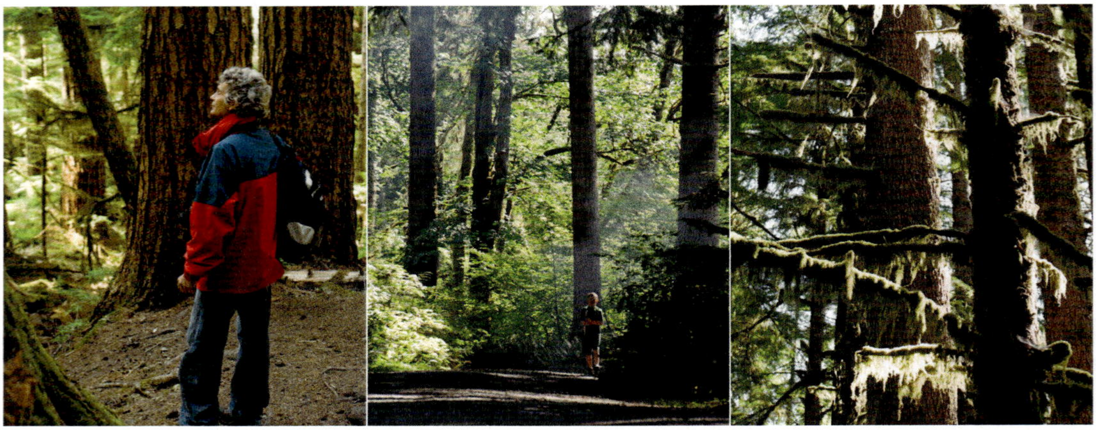

Figures 7-15, 7-16 and 7-17. *Tree worshipper, boy in sun beam and Hoh Rainforest moss.*

Figure 7-15 was shot with a -.3 exposure correction. Figure 7-16 isn't that high a contrast, but it was important not to let the lighted areas on the path blow out. I used the path for a custom white balance. Figure 7-17 is a very high contrast shot. I kept trying to not blow out the backlit moss, but soon realized that it WAS blown-out by nature. Then I concentrated on making sure there was detail in the tree trunks, which was a more successful approach.

SHOOTING IN THE FOG

Fog is interesting stuff. It can look very pretty curled around the base of a bridge. It can soften and cushion hard edges. Or it can screw up your shots entirely. It depends on the density and location of the fog and on where the subject of your photo stands in relation to it.

Think of fog as a lighting tool, not as the subject of your photo. The series of shots below show some of the things fog can do for you.

It can isolate your subjects:

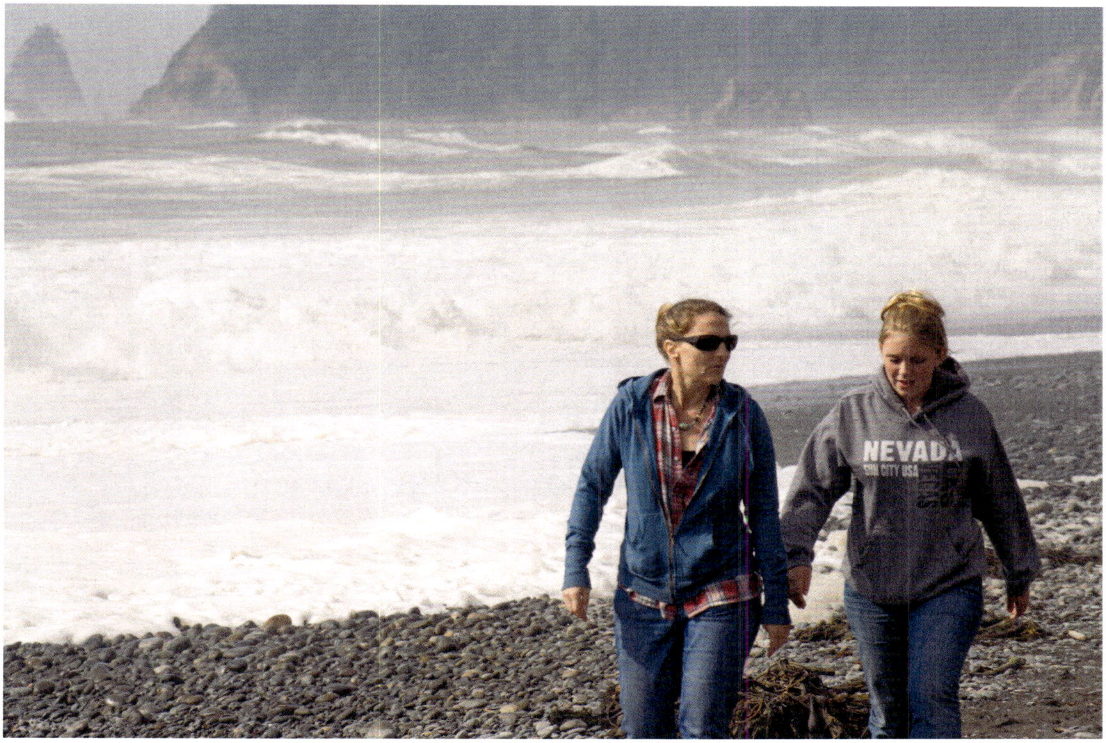

Figure 7-18. *Rialto Beach WA*

It can create mystery:

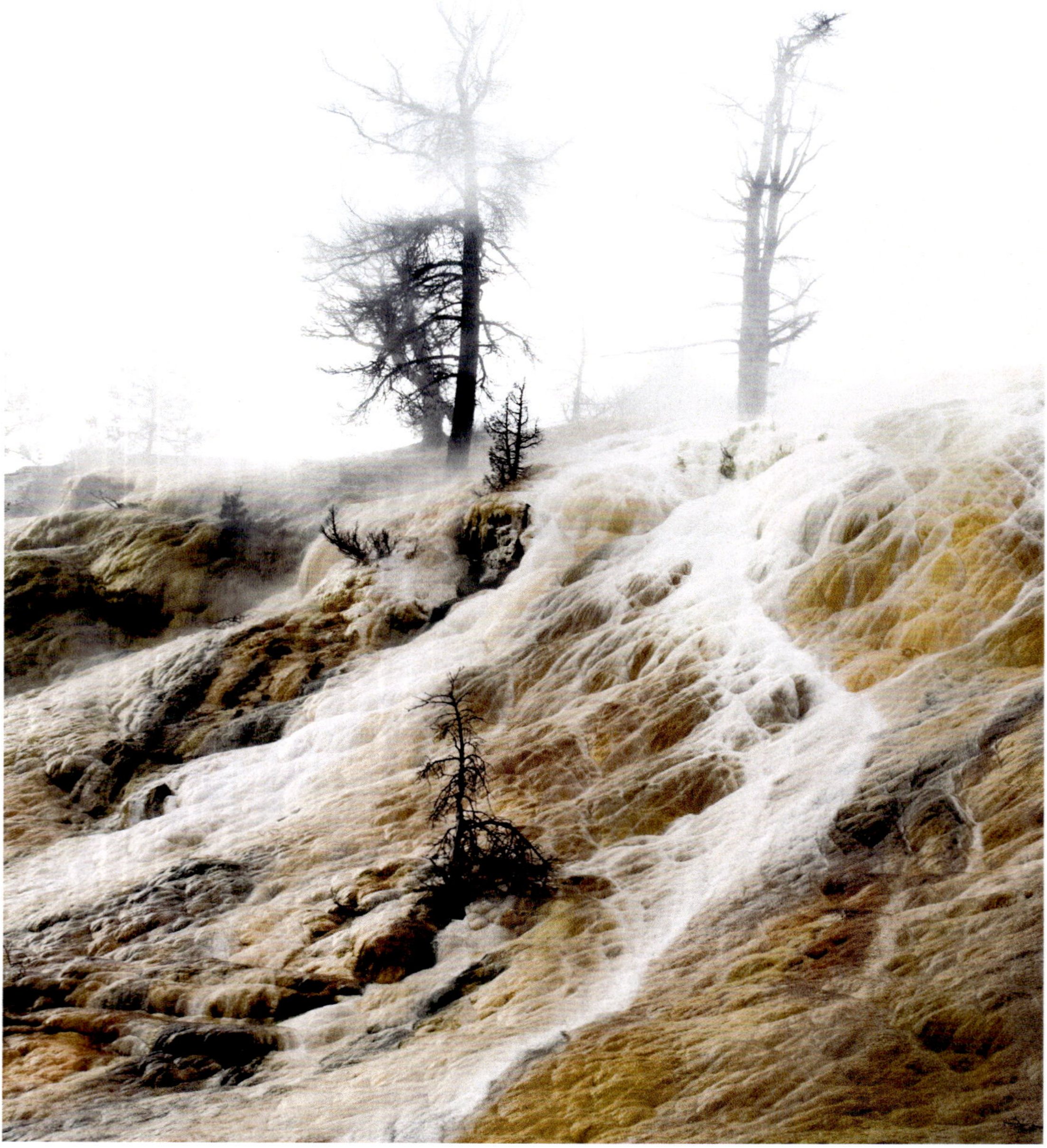

Figure 7-19. *Yellowstone Nat'l Park*

It can frame your subject:

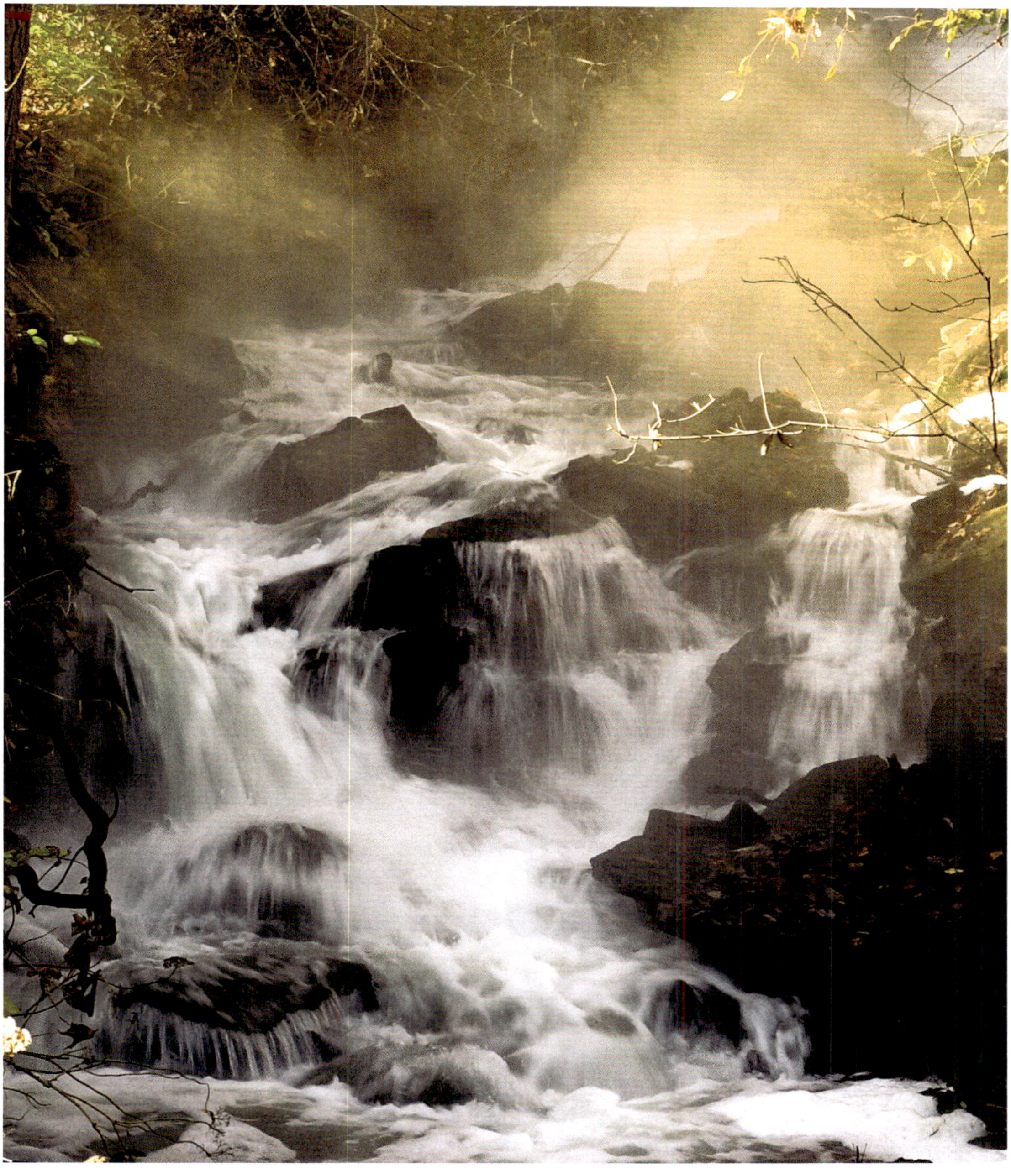

Figure 7-20. *Socialville-Foster Rd, Maineville, OH*

Or it can romanticize your subject:

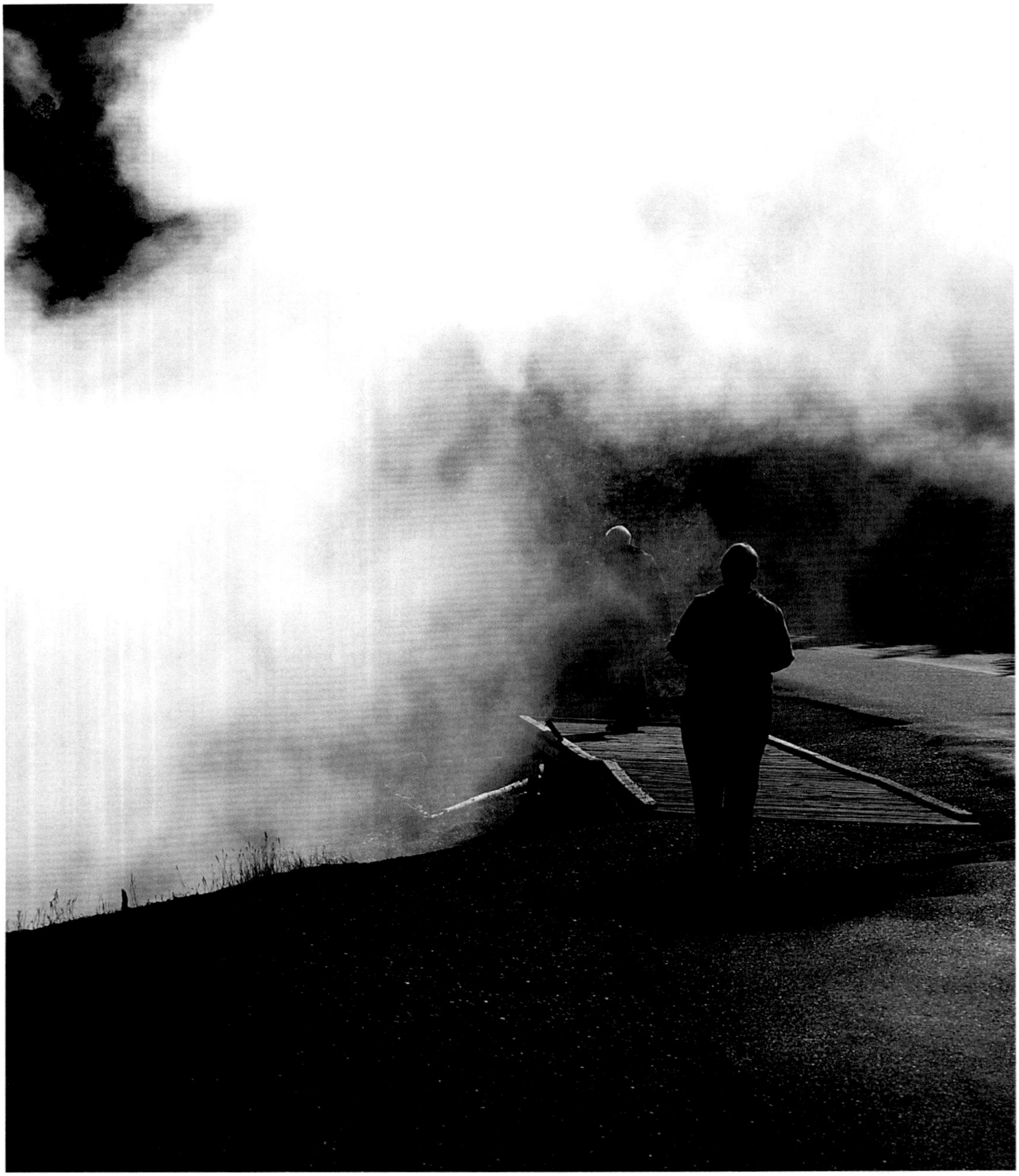

Figure 7-21. *Man and Fumarole, Yellowstone National Park*

Fog tends to wash out color as well as soften light.

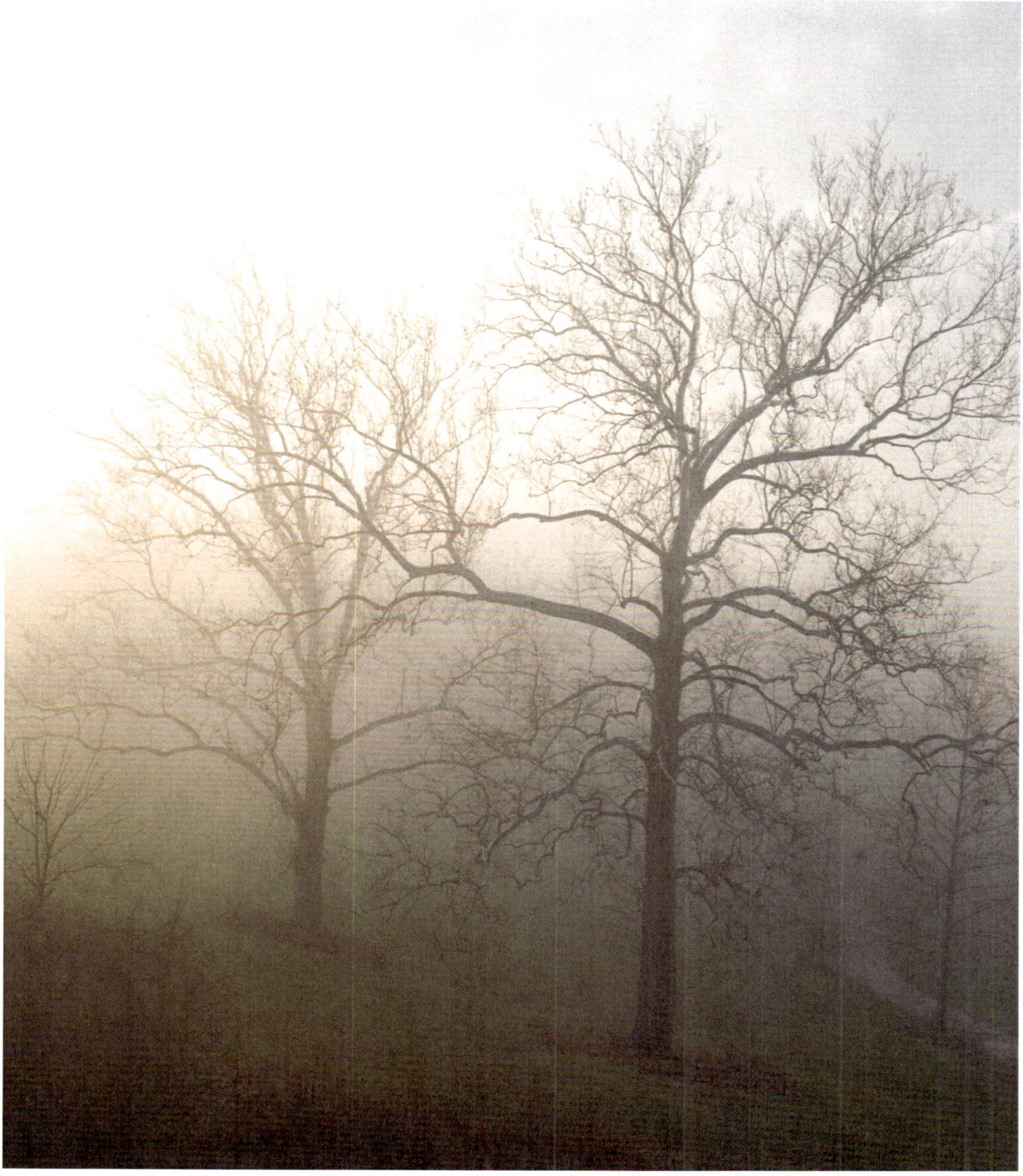

Figure 7-22. *Sycamores in early morning fog, Loveland, OH.*

Here is the same scene (a crop) shot 50 minutes later after the fog had burned off. Color and vibrancy have been restored to the world!

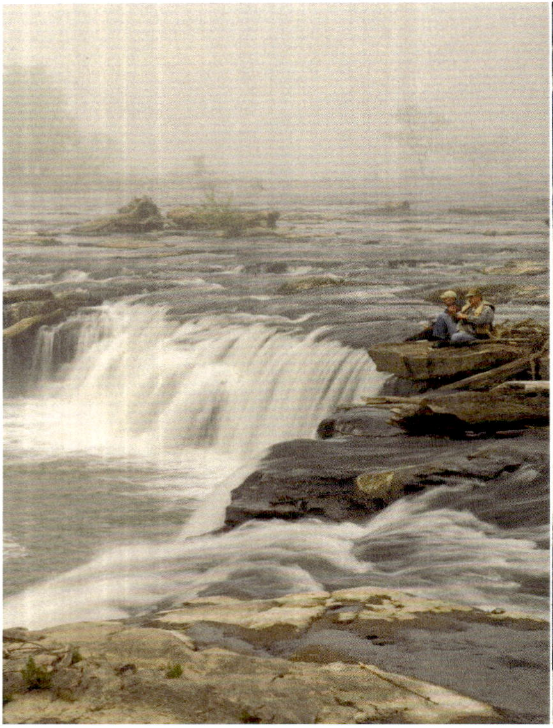

Figure 7-23. *Sandstone Falls, WVA in fog.*

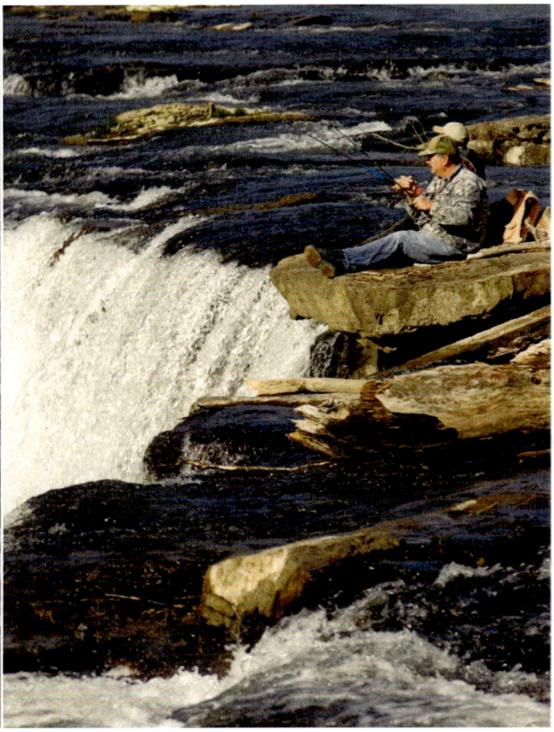

Figure 7-24. *Sandstone falls 50 minutes later. Notice that the same fishermen are present in both shots.*

Now that we've seen some ways to use fog, how do you go about it? It's possible to make fog the subject of your photo, but not if you're in the middle of it!

What works really well is subjects emerging from fog, or disappearing into it. Things too deeply immersed in fog aren't decipherable, and, unless your composition is brilliant, deeply fogged-in objects aren't likely to stand out as your subject of interest.

There aren't a lot of absolute rules for shooting in fog. You are likely to have a low-contrast shot at the best of times; you can expect to do some editing work to bring up black point (enrich the blacks), and some curves or levels adjustment to achieve adequate brightness. WARNING: Be very careful using a "contrast" adjustment. You want more contrast in the picture, but don't overdo it to the point where texture within the fog disappears (if there was any). The photo of the man and the fumarole above was tricky in this regard; the picture wanted, if you can imagine this, crisp fog! I metered on the fog, which meant the picture was dark, but at least there was detail in the clouds. I brightened the shadows by a small amount later to balance the picture.

Tips for shooting in fog

- The foggier it is, the closer to your subject you need to be.
- Use a lens hood. If the fog is thin, you'll reduce dazzle; if it's thick, it will help keep water droplets from condensing onto your lens! Dazzle was almost a problem in the shot of the spring sycamores above.
- Spot meter on your subject, then bracket. You'll have the subject right, and the bracketing lets you work with the balance between subject and fog.
- You can try a polarizer. It may help define the subject a bit without taking away the fog. Remember that polarizers mean an aperture change or a shutter speed change.
- Fog shifts, lifts, curls and swirls. Patience is a huge help in taking fog-bound photos. Just wait until you get what you want, and go make lemonade if you don't get it after all!

USING FILL FLASH OUTDOORS

The use of flash in daylight photography is much less common than it deserves to be. Fill flash can balance shots that would otherwise be turned into mere silhouettes by strong backlighting. It can call attention to the subject (though it won't, properly used, isolate the subject).

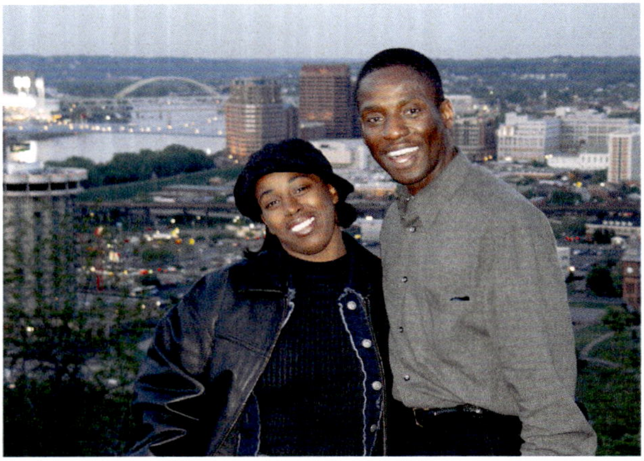

Figure 7-25. *Couple with twilight Cincinnati in background*

Fill flash is a matter of balance and emphasis, not the chief source of light. This shot is borderline; taken at dusk one spring evening, the flash isn't the sole source, but it sure does light up this happy young couple. Use slow-flash or rear-curtain flash to get the most from the background.

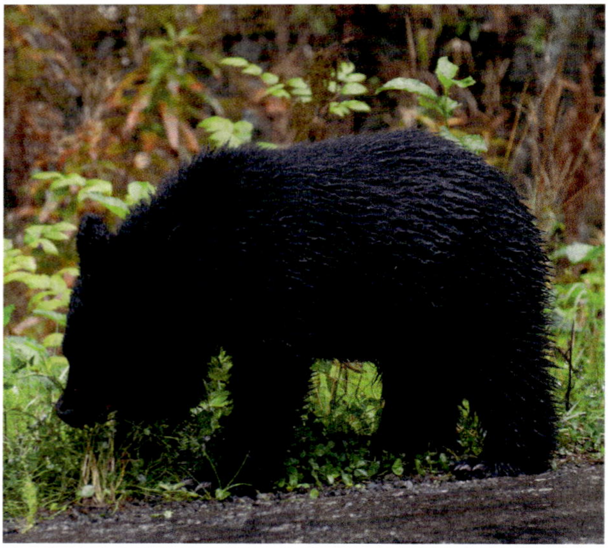

Figure 7-26. *Black bear browsing, Stewart, British Columbia. No fill flash = very murky detail*

What do you do with a very black bear just about sunset in a drenching rain? Fill flash to the rescue! I was driving into Stewart Canada, and saw this fellow browsing with his Mom by the roadside. The first shot was unsatisfactory, with the bear lacking detail and the foliage dominating the shot.

Luckily, I could reach my external flash, rolled down the window, and got three quick flashes before Bruin looked really annoyed. In the next shot you can see much more detail and the catch lights from the flash in Br'er B'ar's eyes.

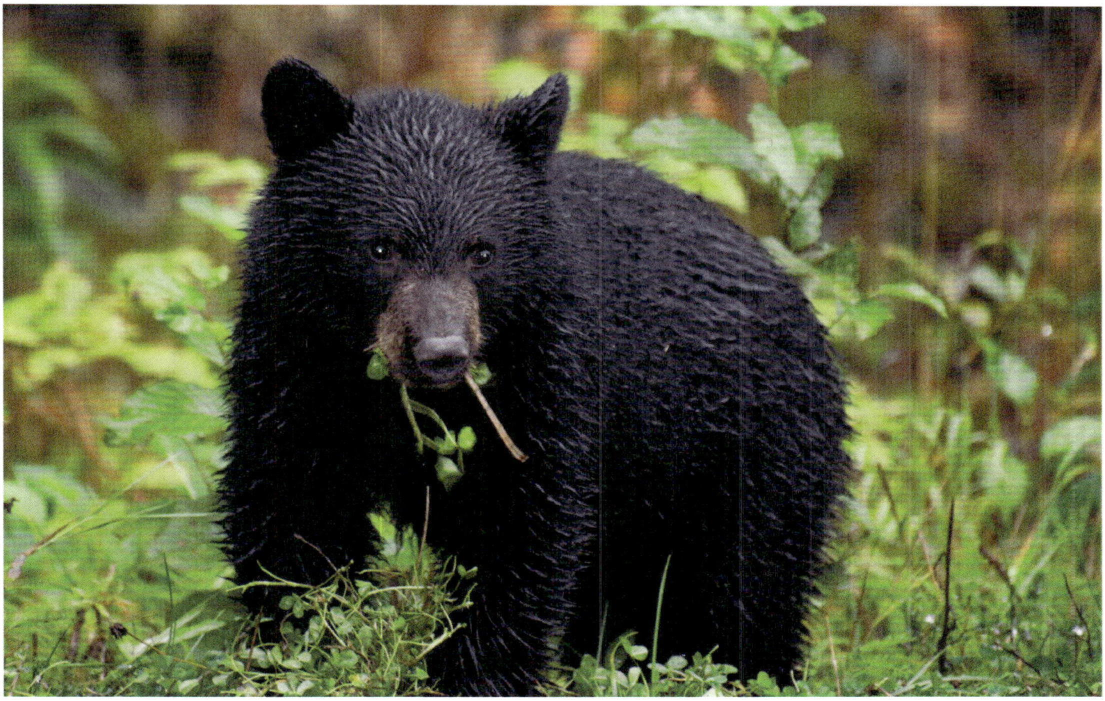

Figure 7-27. *Same scene as 7-26, this time with fill flash.*

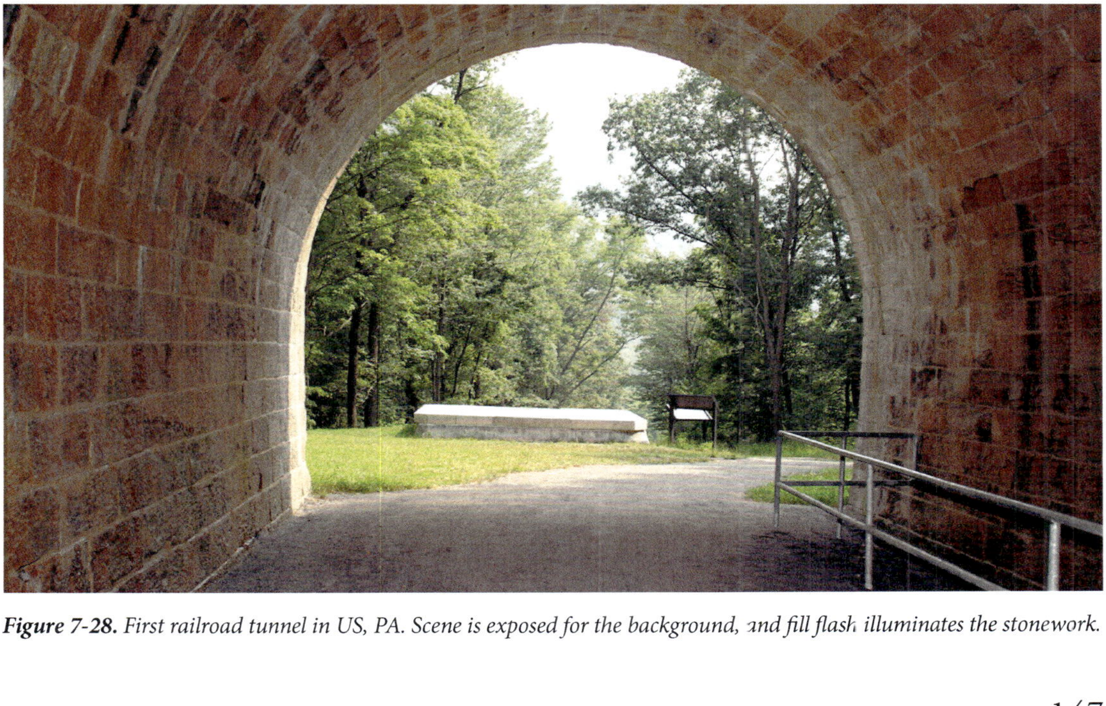

Figure 7-28. *First railroad tunnel in US, PA. Scene is exposed for the background, and fill flash illuminates the stonework.*

In the previous shot fill flash was essential to lighten the stonework and give depth to the nation's first railraod tunnel, one mile long in Pennsylvania. Its job was to haul canal boats through a mountain!

The way to take this kind of strongly backlit shot is to meter the background (center weighted metering works well in this case). Set the flash at -1/3 stop, or even -2/3 stop as in this shot, and there you are.

Figure7-29 illustrates fill-flash to make the image useable. Without it, these guys looked like Dawn of the Dead zombies – no faces, just black. Fill flash solved the problem – luckily I had my external flash mounted for other shots I'd taken a few minutes earlier in another location.

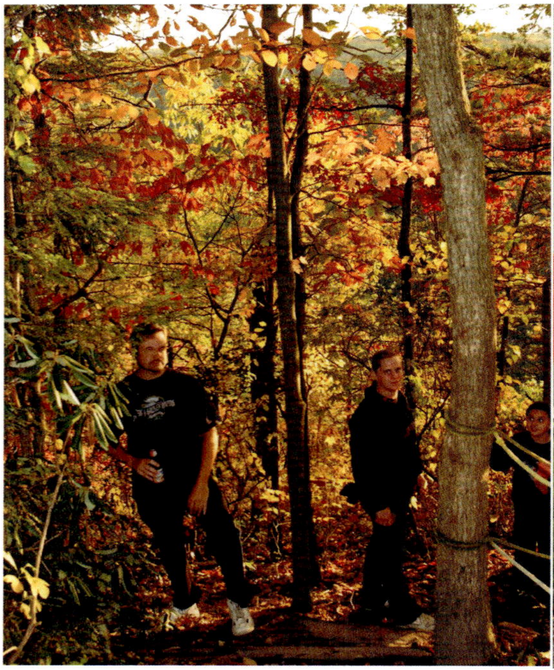

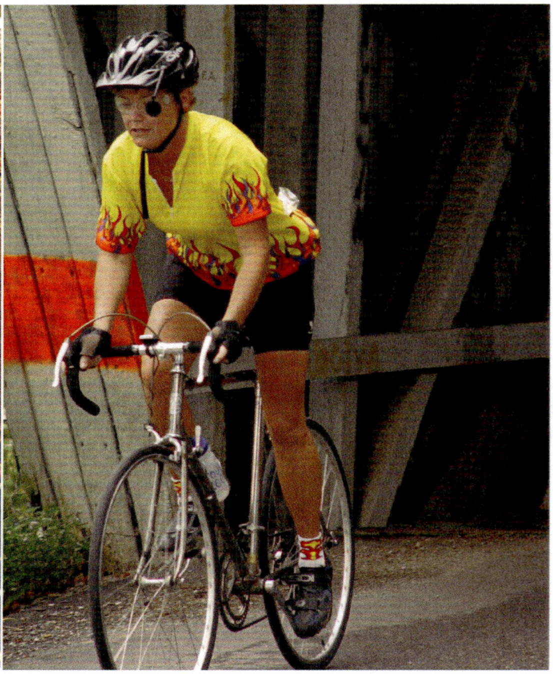

Figure 7-29. *Fill flash is essential for this backlit shot. Class VI, Fayetteveille, WVA.*

Figure 7-30. *bicyclist in Adams County, OH.*

In Figure 7-30, the fill flash does two jobs. It fills in some of the very shadowy tunnel, and it emphasizes the rider by highlighting her yellow jersey.

Mako Haggerty runs a water taxi that operates on Kachemak Bay out of Alaska. Mako had just finished a most interesting interview, and I wanted an environmental portrait. Here's a before–and-after showing what happened when we tipped his hat back a fraction and added fill flash. Composition is improved as well, and the rigging no longer grows through his head.

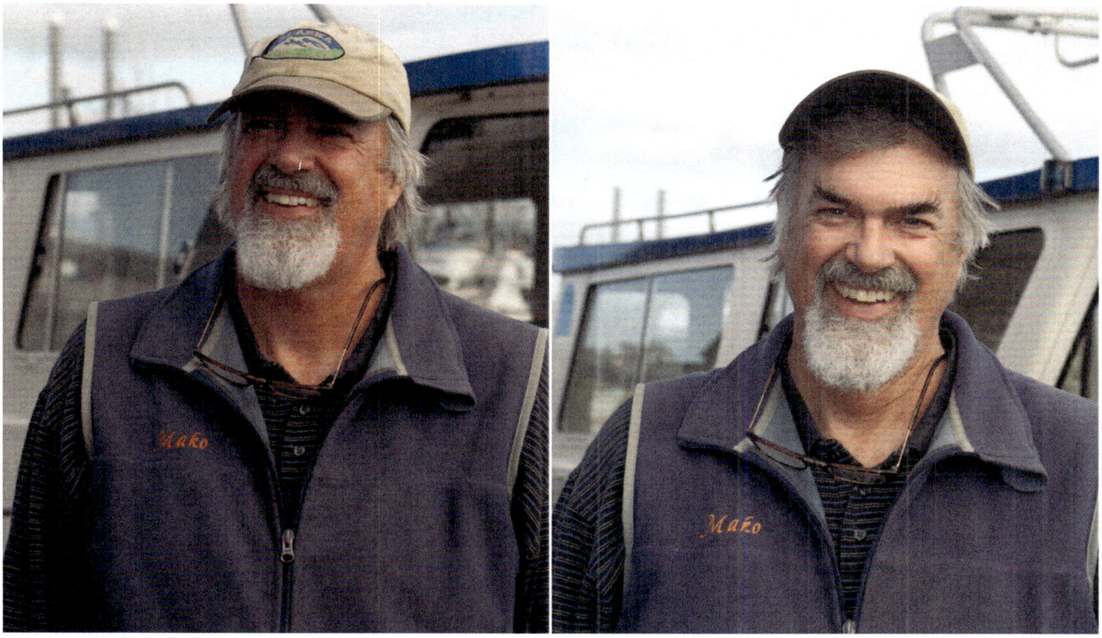

Figures 7-31 and 7-32. *Portraits of Mako Haggerty, Homer, AK. Fill flash makes the difference!*

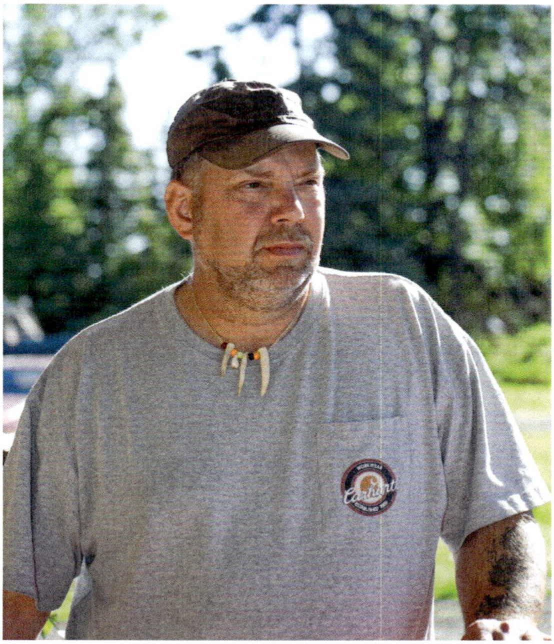

Figure 7-33. *Mike Chapman, Ninilchik, AK*

Finally, this portrait of Mike Chapman, gravel hauler, innkeeper, and bush pilot catches much of his honesty – and worry. Times are tough in Alaska. I don't remember what's causing the color artifacts in the lower quarter of the picture, but the fill flash works pretty well to rescue his skin color and fill in the heavy shadows under his cap. Notice the shadow line on the inside of his left forearm; without the fill flash the whole shot would have been that dark. It's a question, too, of balancing foreground and background.

These pictures are intended to illustrate the various services fill flash can perform for you. To repeat the steps, meter the background, set the camera to that value, and add flash at 1/3 to 2/3 under the metered value.

Nothing to it but to do it— and keep practicing.

SHOOTING IN THE DARK

Shooting after dark poses several challenges. The first one is that, gee, it can be hard to see what you're doing! A small LED flashlight that you can hold in your teeth while you fiddle with camera and tripod is great; so is an elastic-mounted LED to go on your forehead (mine even has red bulbs as well as white to protect night vision; it cost about $12 as I remember).

The second challenge is that you will probably have to shoot in manual mode AND use manual focus ... it depends on how much contrast there is in the shot. You may be able to shoot cityscapes using automatic functions, but a waterfall after sunset – nope.

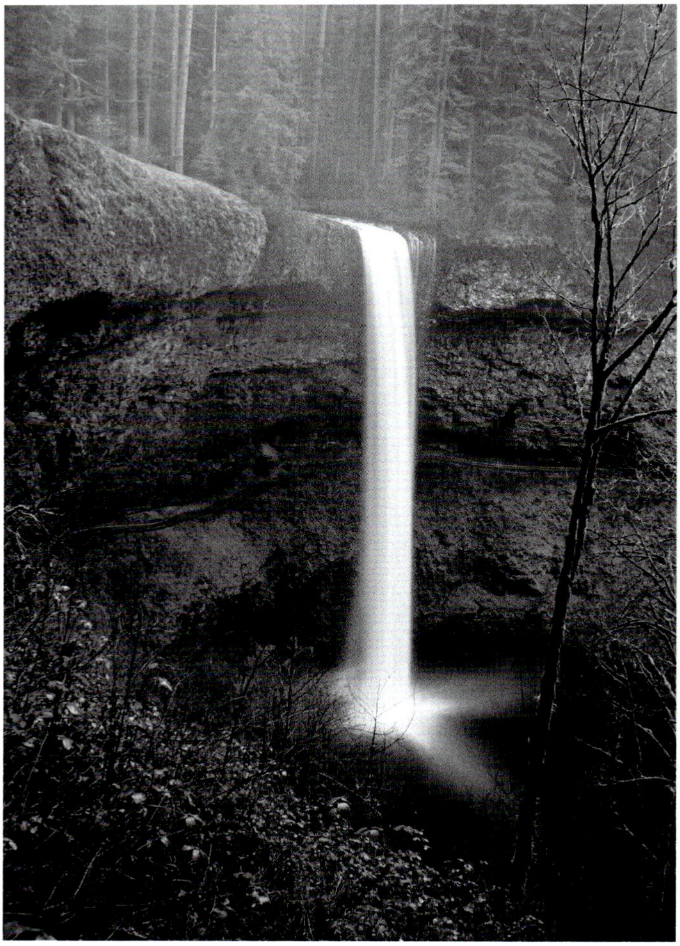

Figure 7-34. *Silver Falls, OR*

Speaking of waterfalls after dark, Figure 7-34 is a shot taken when it was way too dark to read. It was a 30 second exposure, ISO 25. That accounts for the incredible pattern of foam at the bottom of the falls. Had to be done manually, of course. The mist was from the 177-foot fall

Your DSLR camera has a shutter speed setting labeled "bulb." In pre-digital days, one form of shutter release used a rubber tube with a plunger at one end and a squeeze bulb at the other. You squeezed the bulb to extend the plunger, which depressed the shutter release on the camera. As long as you held the bulb, the shutter stayed open. So an exposure of an indefinite length is called a "bulb" exposure. Different cameras will shoot to different pre-set lengths, but pre-sets of longer than 30 seconds are rare; after that it's bulb time.

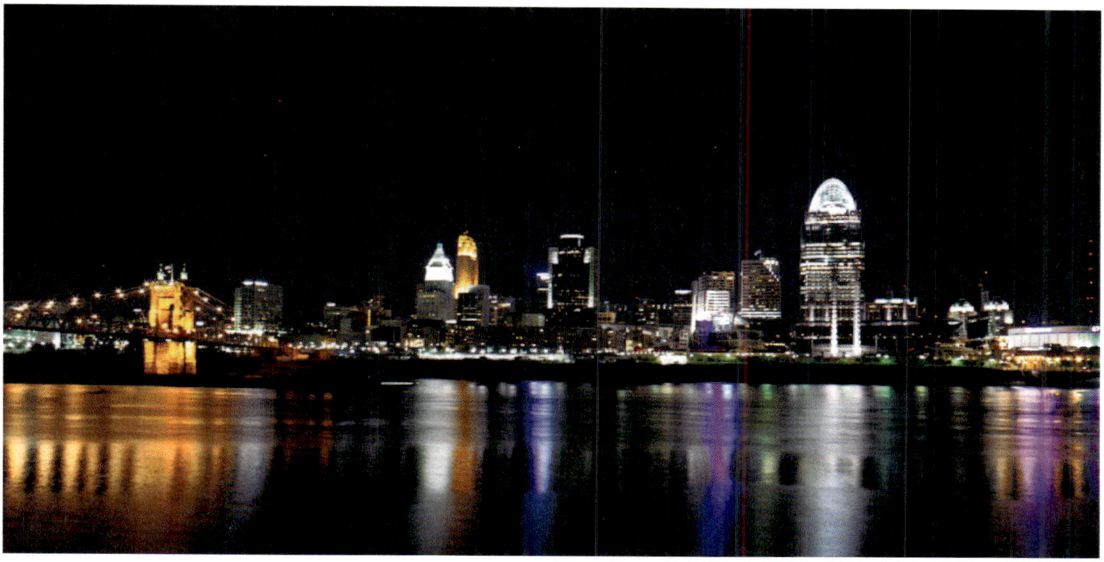

Figure 7-35. *Cincinnati by Night exposure is 6 seconds, ISO 100, f8*

In doing night shots, I like to use the camera's aperture priority mode and take a test shot. If it looks halfway decent, I'll then transfer those settings to manual mode and play from there.

The reason for this is that I don't want the camera changing its mind or trying to compensate by overexposing. If it's a cityscape, I WANT dark darks and light lights; I don't want the camera trying to fill in detail in between resulting in an overexposed picture with grey skies, not black ones.

Some specific tips:

Shooting the Moon: Use manual mode f8, ISO 200, 1/125 second and bracket. The moon is a pretty bright object. Long exposures (over a second) will blur as the moon moves through the sky

Shooting Fireworks: Use manual mode. Use f8 or f11, ISO 200. Use the bulb setting and a remote shutter release – and, of course, a tripod. Pre-focus at infinity or a tiny bit this side of infinity. (Test some shots to see which is better). Open the shutter as soon as the rocket starts up and leave it open until the firework dims out. Use a wide-angle lens; 35 mm is fine. At this point you can try a couple of things. You can try starting earlier and ending the shot later. You can also try using the "multiple exposure" option on your camera and try to stack succeeding bursts together. (Since you're using a tripod, you can also do this easily in Photoshop.)

Stars and Aurora: Shoot wide-open using a moderate wide-angle lens (50mm normal lens is OK, too). Manual focus to a tiny amount this side of infinity. ISO 400 to 800 or even faster is OK, if your camera has good noise reduction software. The higher the ISO, the noisier your picture — but the shorter your exposure time. Try using the bulb setting, using a remote shutter release, and shooting for 30 seconds to 2 minutes. As an alternative, you can try holding the shutter open for half an hour at ISO 100. You'll see star tracks, but possibly a meteor, interesting tracks from passing planes, and so forth.

For aurora, same as above but keep the duration as short as possible because the aurora are always shifting and will blur out on you. F8 for 15 seconds is a good starting point. For both stars and aurora, shots with some horizon features (trees, for instance, or a lake) help give scale and a sense of wonder to your pictures.

CHAPTER 8: *Specific Situations*

SHOOTING GRADUATIONS
SHOOTING IN AQUARIUMS
SHOOTING MOTORCYCLE RACES (AND OTHER FAST MOVING EVENTS)
SHOOTING PEOPLE PICTURES OUTDOORS
SHOOTING WATERFALLS

CHAPTER 8: *Specific Situations*

SHOOTING GRADUATIONS

It doesn't matter whether you're being paid, or whether you're shooting your niece's graduation as a gift; either way there are just four "money shots" you HAVE to get, and the rest is gravy.

What counts in all these shots is emotion and relationship. Get those, and everyone will love your work, even if it's a tad shaky on the technical or artistic side.

1. "The hand-off." This is the hardest to get because you get one crack at it, and the moment is fleeting indeed – well under ten seconds is the norm. There are 700 graduating seniors, and the speeches were long enough — the deans or principals can't wait to get it done! Indoors or out, you set up for this one the same. Here's how:

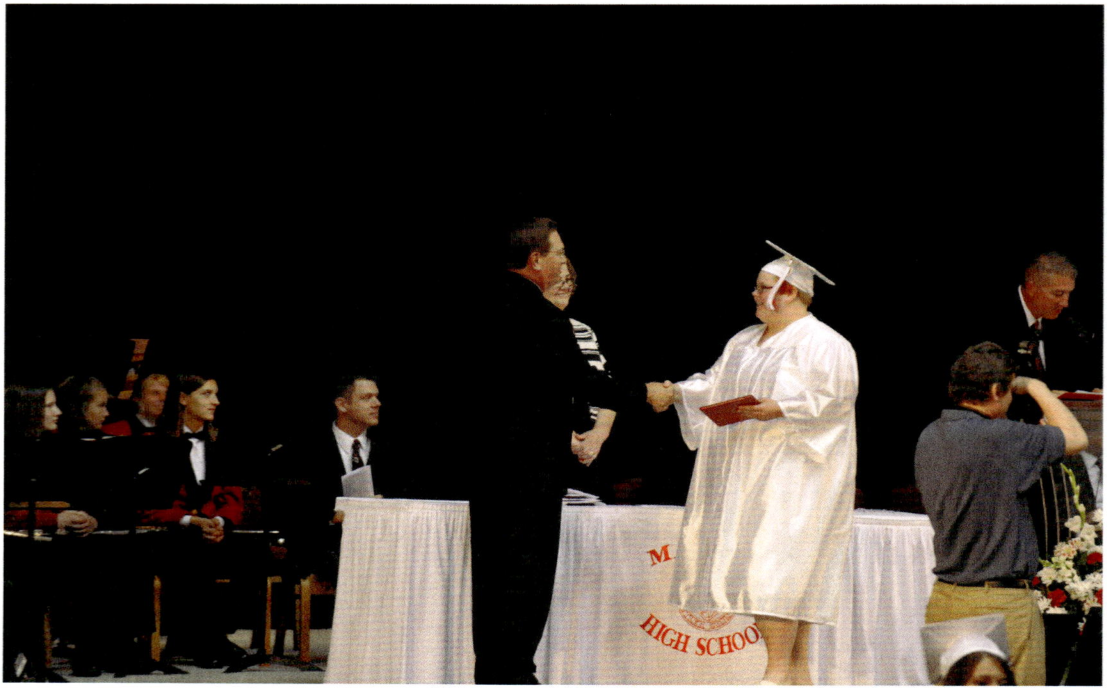

Figure 8-1. *This is THE money shot – getting the Diploma (The Hand-off).*

Get there early. So early you get your choice of seat with a great view, as close to the stage as possible, but preferably high enough up so that you won't be shooting way up in the air to get the shot. Not flattering, that one.

Have your equipment ready, your exposure values tested, your long lens in place, batteries renewed, lens and sensor cleaned, etc.

Take practice shots of people you don't care about. There's a rhythm to this shot, because there's a rhythm to the event itself. Get in the swing of it.

If your camera will do it, switch to continuous autofocus and rapid fire. This may give you more than one bite at the apple, so even if you don't get the perfect shot, you'll get some shot of the hand-off.

2. "The cap toss." Just what it says it is. Another shot where rapid fire is really useful. But be careful – don't get so excited doing this that the camera gets jumpy. Pre-focus and shoot in manual focus, or trust your autofocus. Figure 8-2 was taken when it was really dark (after sunset) and the heavens were ready to burst. Shot at f3.5, 1/250, ISO 2000, 28-70 lens at 44 mm.

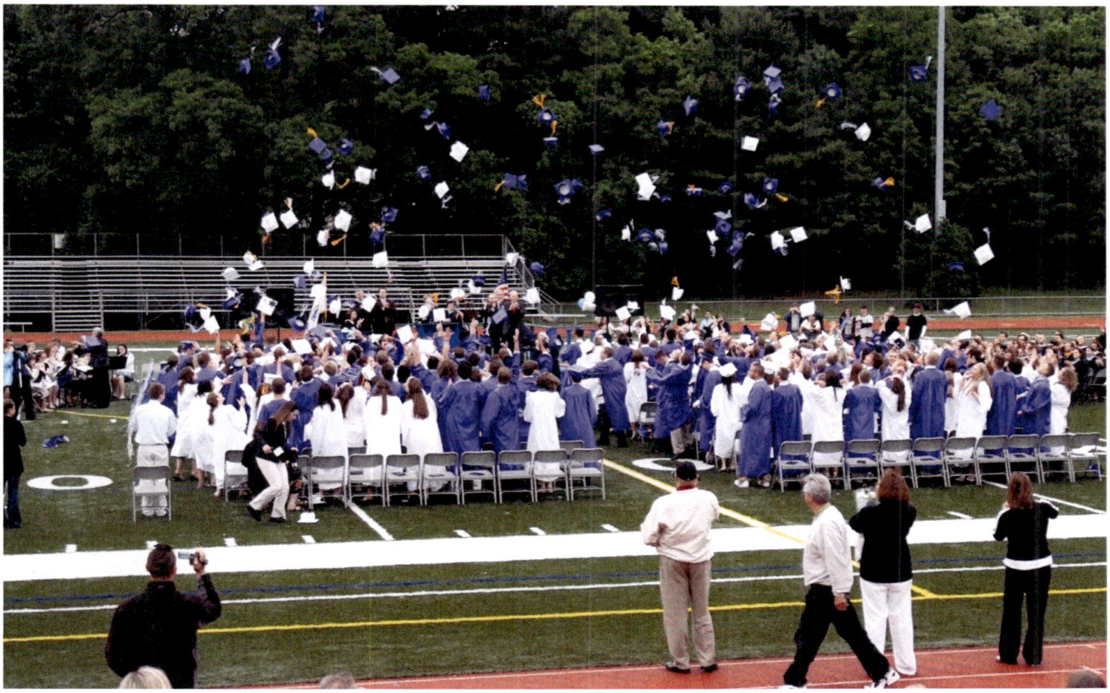

Figure 8-2. *The cap toss.*

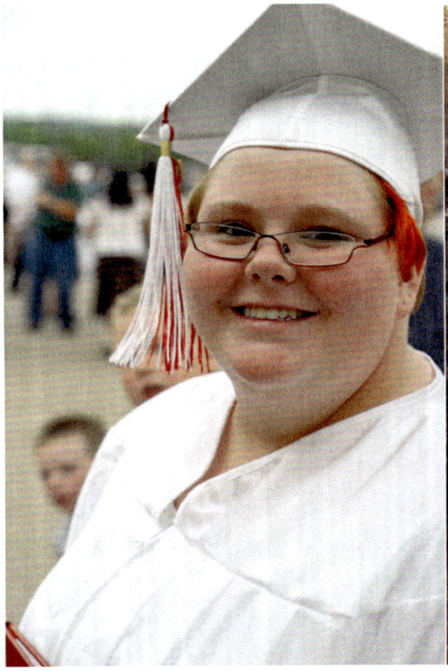

Figure 8-3. The happy grad.

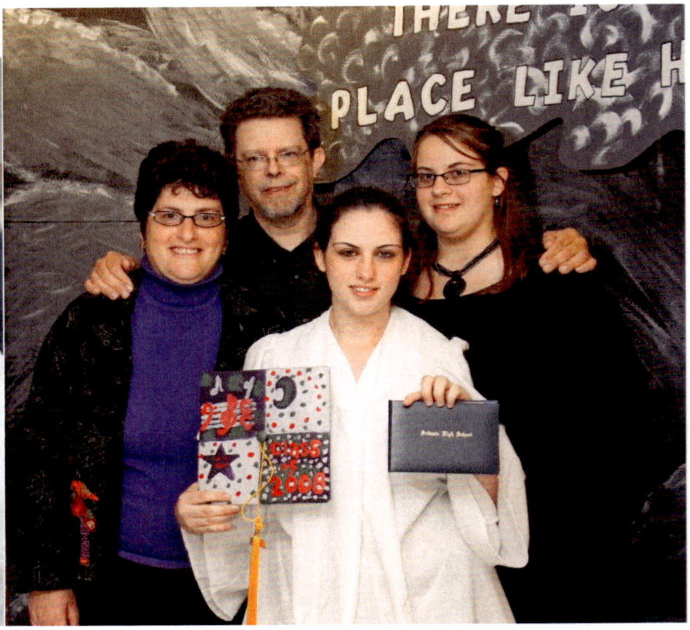

Figure 8-3. The happy family.

3. "The happy grad." You can take this with or without the diploma, but he or she MUST be wearing the gown. Cap is good too, but it may be sailing for France by the time the ceremony's over. You can take this before or after the actual graduation, but right afterwards gives you the most spontaneous look.

4. "The family." Here you want to show closeness and happiness. Posing is really important here, and you will probably want to take more shots than your subjects wanting to take. Don't push it; scowls aren't what you're after. Technically, these shots are no different than other people shots.

There are plenty of other graduation shots, which are nice-to-have. Classmates, dignitaries, hugs and squeals, beaming relatives, band players, after party —all are valuable for the graduate's album or CD. But these four are the make-or-break shots that make your graduation shots a hit.

SHOOTING IN AQUARIUMS

Aquariums have gotten better and better in the past few decades. Boston, Baltimore, and Newport KY (across the river from Cincinnati) have built wonderful facilities. I can't even think how expensive it is to stock and maintain those walk-through tanks and amazing habitats!

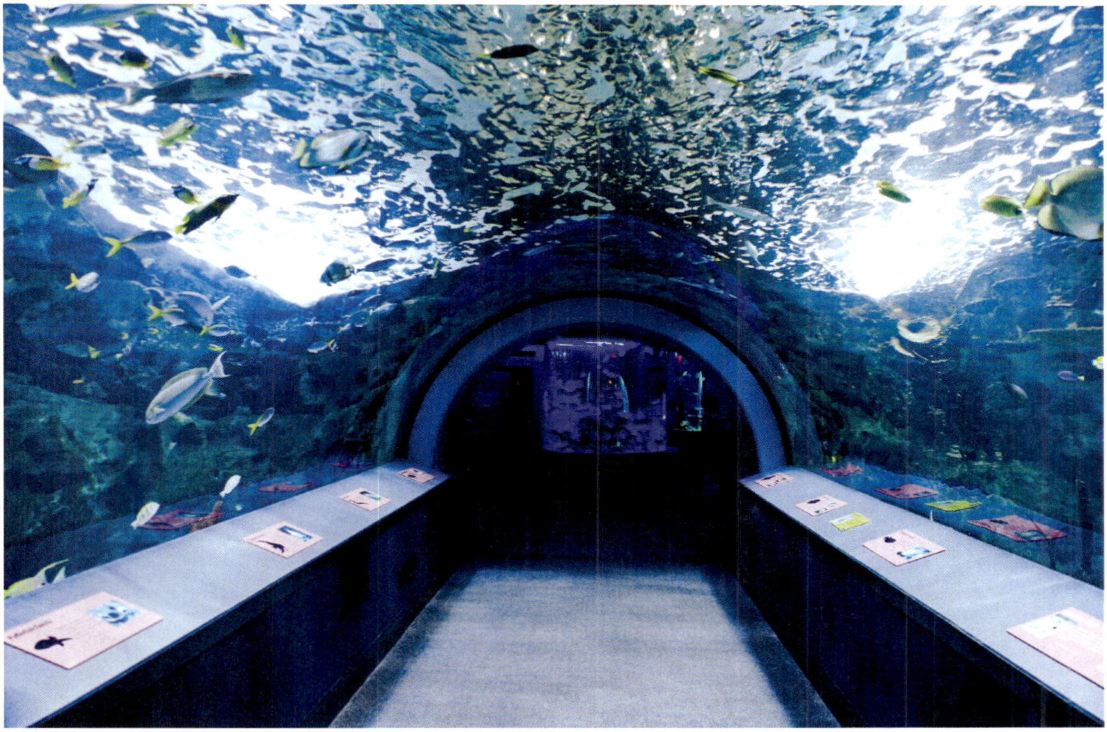

Figure 8-5. *Glass Tunnel at Newport Aquarium.*

These new aquariums are a magnet for photographers. But they're not easy places to photograph for a combination of reasons.

The creatures in the aquarium are fascinating, and the Newport Aquarium, where these shots were taken, has tried hard to keep the water really clear and to provide interesting side tanks with good lighting. But the lighting varies from dim to dimmer, and color fidelity is very difficult since water sucks red out of the spectrum. Additionally the subjects are moving. Tripods aren't permitted without special arrangements.

Unfortunately, the glass / plastic windows compound the problem. The glass can be somewhat dirty on the inside, and on the outside where sticky little fingers take their toll. Distortion from the thick curved plastic tubes you walk through creates a tendency towards blurred edges and chromatic aberration at the edges as well. The straighter you shoot, the better results; angling upward or downward through the glass invites trouble. Here's an example:

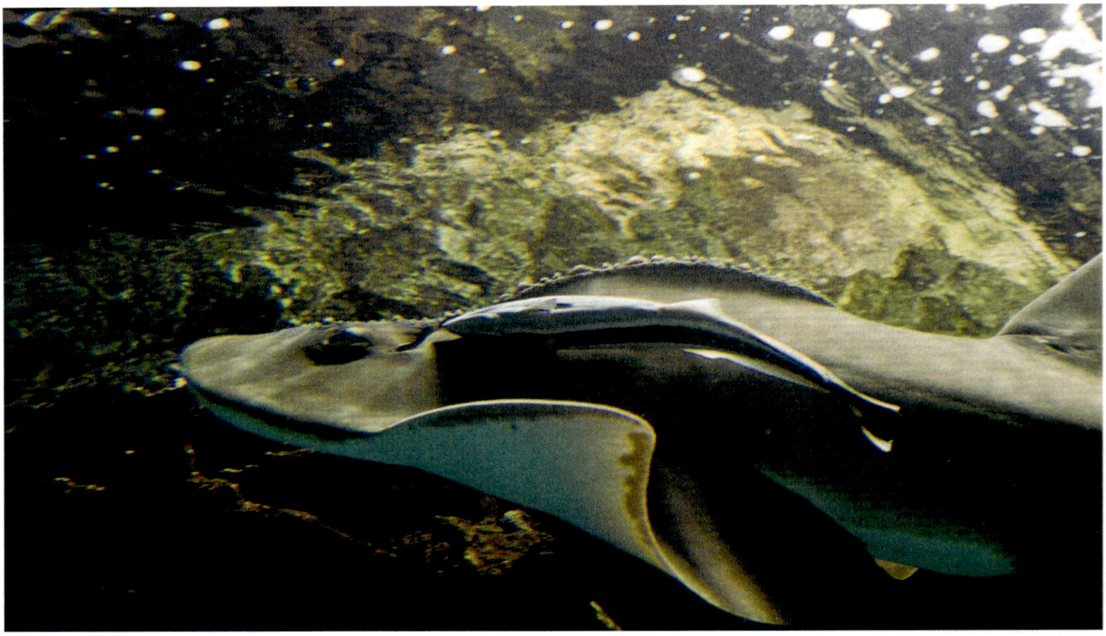

Figure 8-6. *Skate Shark, Newport Aquarium KY*

You can see the smeary quality, especially on the edges of the skate shark's fins. This problem is more visible in larger size prints.

So here's the challenge: you need to get the shutter speed up. This means very wide apertures, which in turn means limited depth of field. The first time I tried this I had a camera that was very noisy over ISO 400. More recently I used the Nikon D700, which starts getting noisy over ISO 3500, but is acceptable above that depending on conditions. This is the one time I'd recommend a DX (digital camera size) sensor over an FX sensor (full frame 35mm), because it has a greater depth of field at any given aperture.

Most of my aquarium shots were taken at ISO 3500 - 4000, f3.5 to 5.6, with white balance varying from settings around 7500 K to auto. No flash. Shutter speeds varied from 1/30 to 1/350. What you really can't do is set-it-and-forget-it because conditions continually change.

A word about flash. Tourists use flash all the time, and what they get when they go home is a very bright blob of reflected flashbulb and a murky-to-black tank with maybe a tiny bit of critter emerging into view. I can't think how disappointing this must be to many, many families.

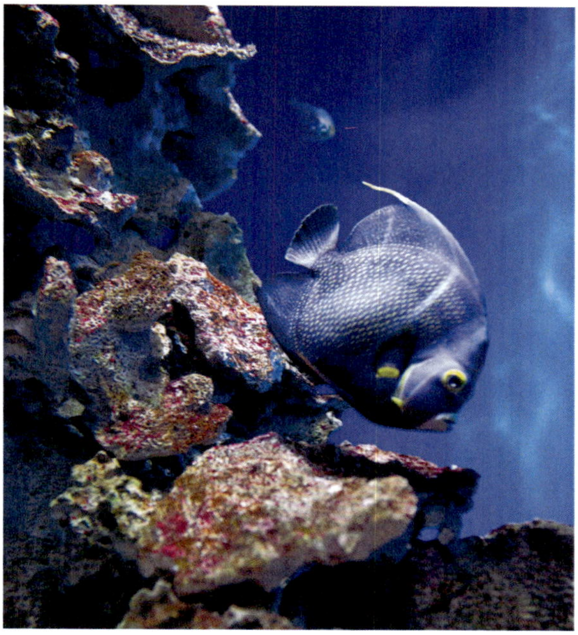

Figure 8-7. A small tank holds a single fish; shallow depth of tank helps with clarity and color of image, but DOF remains an issue.

If you could set up good lights, of course there's a way to make them work. But for all practical purposes you have to shoot in available light.

The smaller the tank and the clearer the water, the better shot, as in Figure 8-7.

Even in this shot, shot at f3.2, ISO 4000, 1/100 second, you can see how the shallow depth of field blurs the rock in the foreground and is falling way off a few inches past the fish's tail. In this particular shot, the shallow DOF is actually a benefit, helping to isolate the subject of interest.

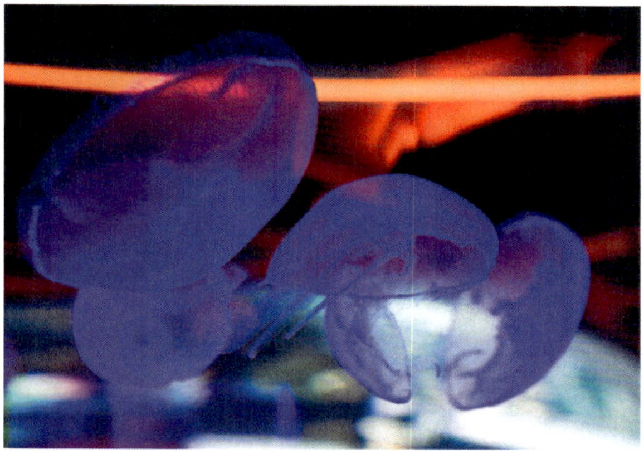

***Figure 8-8.** "Moon Jellies," Newport Aquarium*

The jellyfish shown in Figure 8-8 was an interesting and very rewarding shot to take. The animals were displayed in a 4-foot (or so) diameter glass column of water; the very jazzy lighting was from colored bulbs mounted around the perimeter of the column about 8 feet up. The moon jellies were shot at ISO 4000, f3.2, 1/40 of a second. I've removed some noise from the image, but otherwise it's close to as-shot. The photo is in focus, but these creatures are 99% water, so they always look fuzzy around the images.

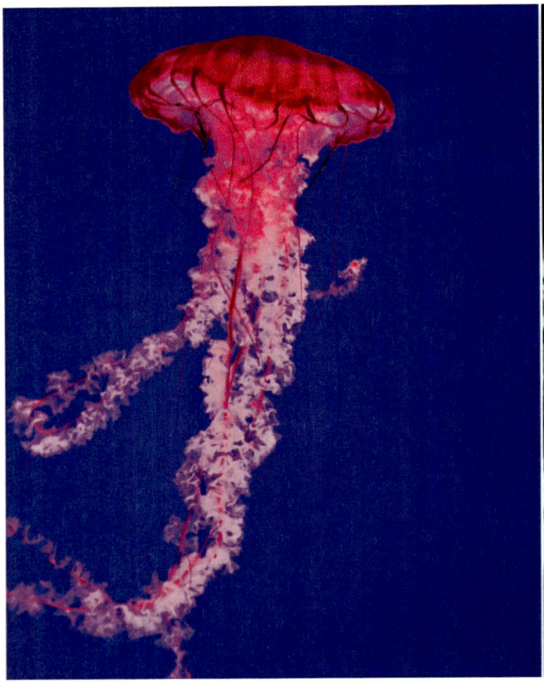

Figure 8-9. *Western Nettle. Newport Aquarium, KY.*

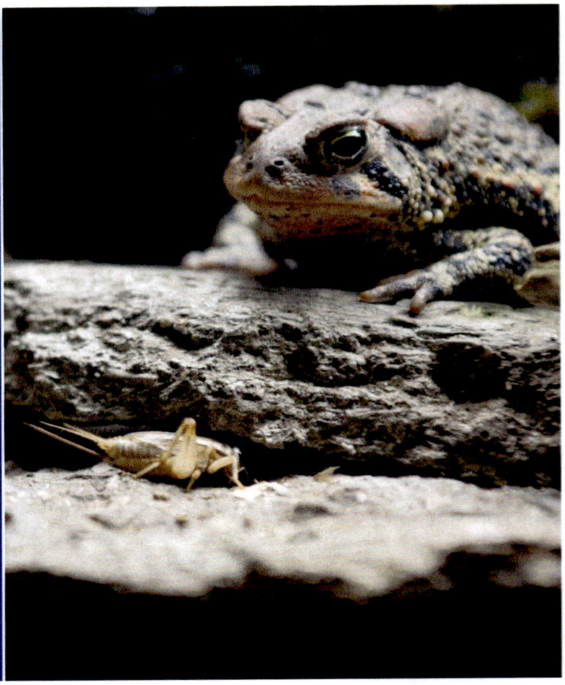

Figure 8-10 *American Toad, Newport Aquarium.*

The Western Nettle jelly takes its color from the Aquarium's lighting. Same exposure as the moon jellies above, with a -1/3 stop compensation. This image also succeeds as a high-contrast black and white.

This aquarium has a number of quasi-aquatic specimens. The American Toad is my favorite of these; pale crickets are waiting to be lunch. I didn't see one being caught, but the Toad's dyspeptic smirk makes me smile. Note that the glass isn't thick, so it doesn't smear the image, but the low available light makes the DOF problem persist 1/60th, f5.6, ISO 1600, -1/3 compensation.

To sum up, aquarium shots require fast shutter speeds in lousy light, since tripods are usually frowned upon and the creatures are moving. But the price you pay is wide-open apertures and high ISO (sensitivity) settings … leading to noise and shallow depth of field.

Many aquariums will arrange "photographer days" or club outings, and at these tripods are OK.

Take your shots when the creatures are as close to the glass as possible and as straight-on as possible. Enjoy the challenge!

SHOOTING MOTORCYCLE RACES (AND OTHER HIGH SPEED EVENTS)

Mid-Ohio aceway features a twisty, hilly course, around which motorcycles zoom at speeds to 180 miles per hour. Watching the motorcycles zoom around the twisty, hilly course gets your heart pumping but capturing the racers is a challenge.

If you ever want practice shooting fast-moving objects in widely varying conditions, this is your chance! I imagine photographing skeet shooting and airplane pylon racing would be harder, but motorcycle racing is hard enough – and a blast.

To do this job you need a fast autofocus system, a very long lens (as fast as possible), and quick reaction time. If your lens isn't particularly fast, you're going to need to shoot at high ISO and hope you don't get too much noise. This is one area where quality glass makes a big difference.

The key to success is panning, where you turn with the action keeping the subject in the center of your viewfinder. This blurs the background while keeping the subject sharp.

Here's what's going on. Imagine you're located at a fixed point. You can spin around, but you can't move away from your spot. Now, imagine a motorcycle located, say, 1/2 mile from you traveling along the edge of a perfect circle. Imagine also that there's a long rope connecting you, so that as the motorcycle travels along, you pivot to keep it in the viewfinder. Finding the subject, framing it, focusing, and pressing the shutter can take four or five seconds, which is all the time in the world. So far, it's easy.

Now imagine that same motorcycle is 250 feet from you, coming towards you, and at 100 feet or so suddenly veers away. Oh, yes, and it's doing 180 miles an hour. Not so easy now. First, your autofocus is going to fight to keep this target in focus because the distance to the camera is changing really fast. Second, you now have about 1/2 second to compose, lock, and shoot.

I was shooting mostly at 1/1250 of a second, ISO 1200, f5.6 with a 70-300 mm lens. A 500 mm would be better, but for a good one you're talking serious money. (I had the loan of a Sigma 50-500 from friend Ric Hartigan for the first of the three shots below).

You'd think that 1/1250 would freeze the action solid — but it doesn't always. In shots where the motorcycle was closest to me, say, 60 feet away, and traveling in a straight line at, say, 130 miles an hour, I'd see motion blur, which you don't want, as compared to background panning blur, which you do want. That's because the motorcycle is traveling about 6 inches during your shutter click (do the math; that's about right). Even if you're perfectly tracking it (pivoting very rapidly and hoping you don't fall over your own feet or bump someone next to you), you're going to have some motion blur. The closer to you the cyclist is running, the more blur you're going to get.

Depth of field is another issue. If the bikes are close to you, moving in a line, and you're shooting wide-open to get the shutter speed up, you're going to have relatively shallow depth of field. The cure for this is to get high up and shoot at a downward angle. Keep the aperture small (over f8) by increasing ISO value way beyond what you'd normally use for a brightly lit scene.

Learning how to anticipate (listen for the engines changing pitch), how to keep the panning motion smooth (follow through; don't stop when you think you've got the shot), how to get the best results.... for me, that was fun.

Enjoy the pictures! You'll get similar training from shooting volleyball, diving, and parachute landings. Hope you have as much fun as I do.

Here are three images illustrating the possibilities.

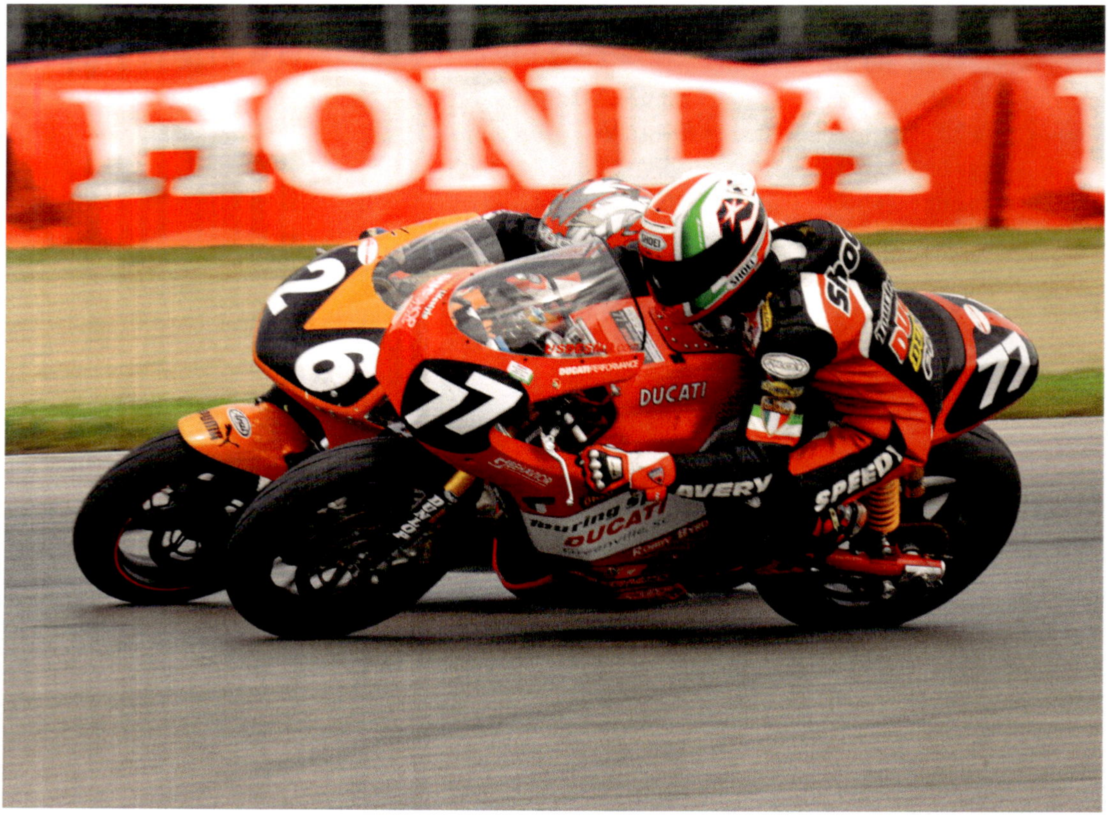

Figure 8-11. *Mid-Ohio motorcycle race, Neck and Neck. F14, 1/640 second, ISO 1000, Sigma 50-500mm at 500.*

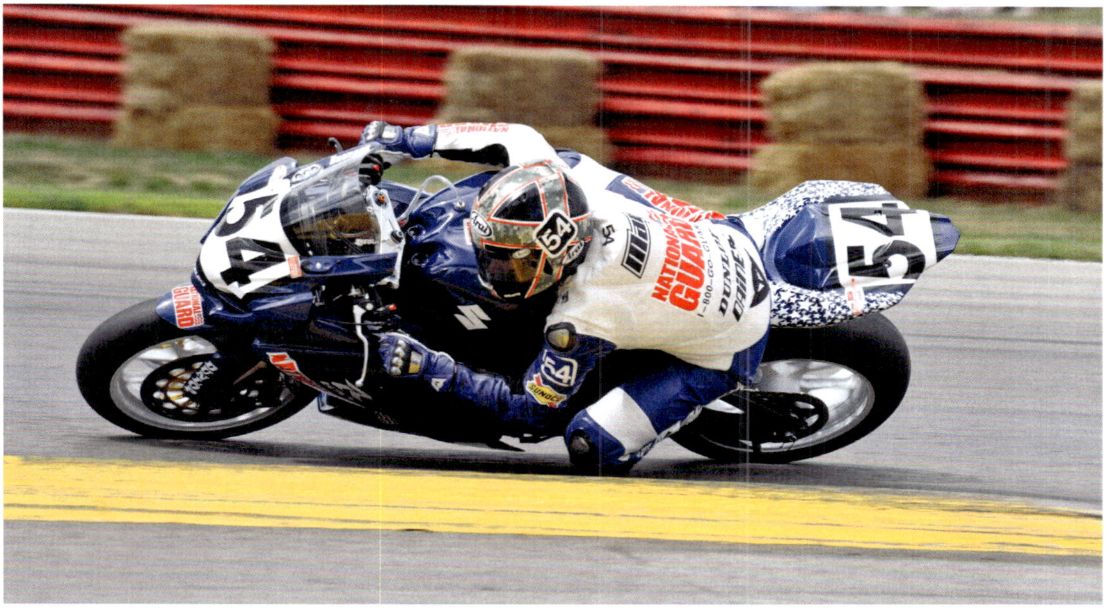

Figure 8-12. *Mid-Ohio. f10, 1/1000, ISO 1000, Nikon 70-300mm at 300mm. Note the motion blur in the background, even at 1/1000 of a second. These babies are moving!.*

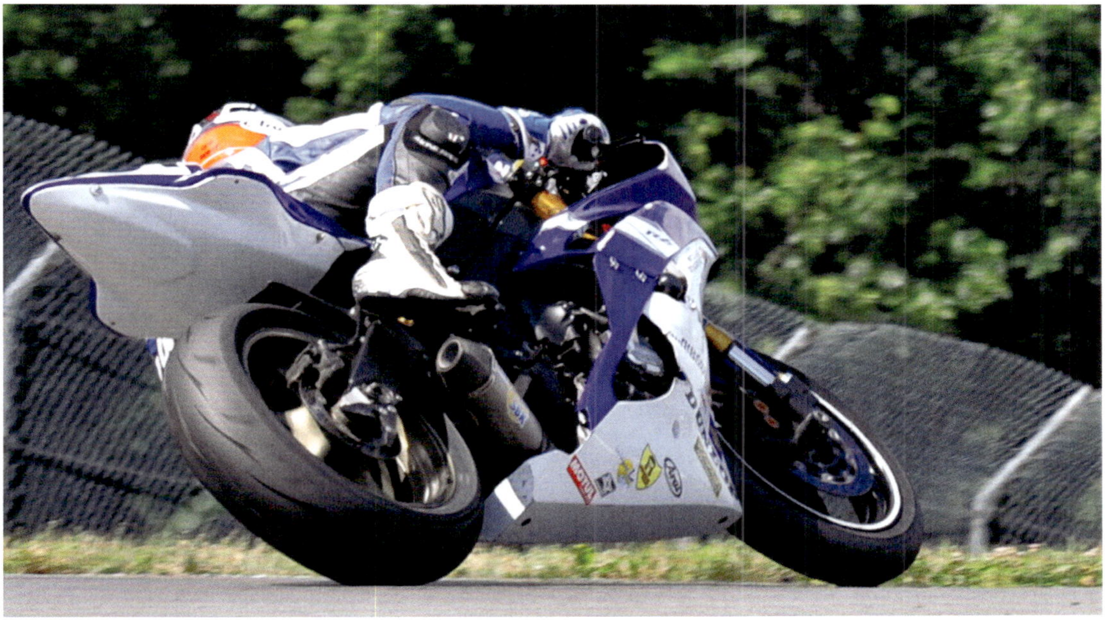

Figure 8-13. *Mid-Ohio Motorcycle Race. F10,1/1600, ISO 1250 -1/3 stop. Extreme low angle adds drama.*

SHOOTING PEOPLE PICTURES OUTDOORS

Shooting pictures of people outdoors is a great American pastime. Lots of amateur and even some pro photographers say, "I shoot outdoors because then I don't have to mess with lights." Well, sorry to say, you ALWAYS have to mess with lights – even if the messing is only to create shadows, reflect sunlight onto your subject, or fill in shadows where they're too dark.

The basic good practices for shooting any portraits apply equally to outdoors work. There are plenty of books, tutorials, webinars, workshops, magazines, and blogs to go through all that. So I'm going to assume you know how to pose families and individuals and will address only the problems directly caused by the outdoor environment.

First: too much sun and sun in the wrong place. A brilliant day is nice to be in, but causes problems for the photographer. First, you'll suffer from too much contrast between brights and shadows. Secondly, if the sun angle is wrong, you will cause your subject to squint, blink at the wrong moment, or with the sun high over head, come down with a bad case of Lurch disease, where the bones over the eyes create dark shadows that are NOT windows onto the soul.

Your goals in very bright conditions should be to create or find light shade, to reduce the very contrasty light/shadow problem, and failing that get your subject to tilt or turn their head so that the eyes are lit but do not squint. If the sun is fairly high but not right overhead, you may be able to place your subject so you're shooting into the sun. Use a gobo (light blocker), the brim of your hat, or even a magazine to shade the lens and use a fill flash if the face is shadowed.

Some examples follow from a workshop I led for the Ohio Valley Camera Club a year or two back. The model was actually a volunteer from the local Chamber of Commerce, and she was great.

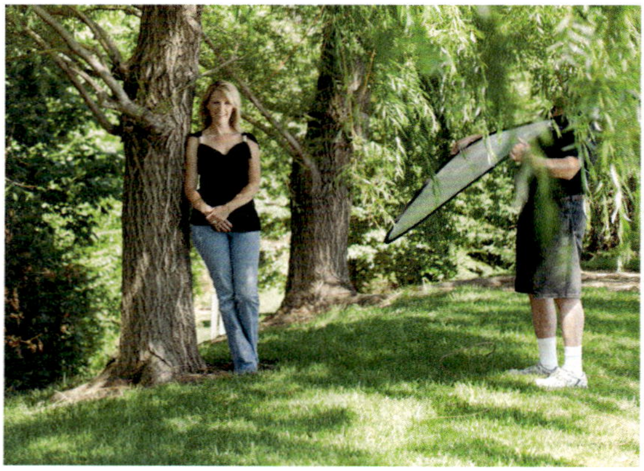
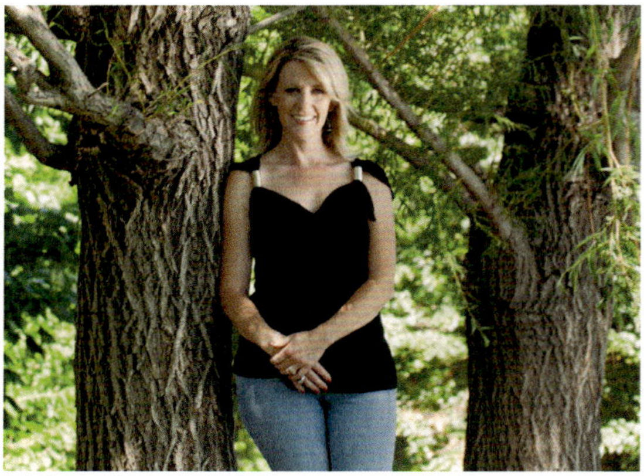

Figures 8-14 and 8-15. *Club Shoot, Maineville, OH. Whole view and cropped. Reflectors are your best friend!*

The left-hand picture shows a reflector being used to throw light onto her left side; the right hand picture shows the result.

Trees produce uneven shadows and bright spots; that's why reflectors and sun-blockers are so helpful in making sure there aren't any glaring hot spots where they should not be.

The other major problem with trees is that they cast a greenish tone onto skin that is very hard to remove. Using a white card or a grey card to set a custom white balance will REALLY help. We don't tend to see the color cast when we're just looking at people, because we see what we expect to see. But the camera sees what's actually there, and if you're going to be shooting under trees and on lawns, you will need to correct for it.

Figures 8-14 and 8-15 are two more frames from the same workshop. The pose is better on the left, but the girl's left eye is too dark. The right one has a less appealing pose, but the eyes are better lit. The difference is a slight change in the position of the reflector.

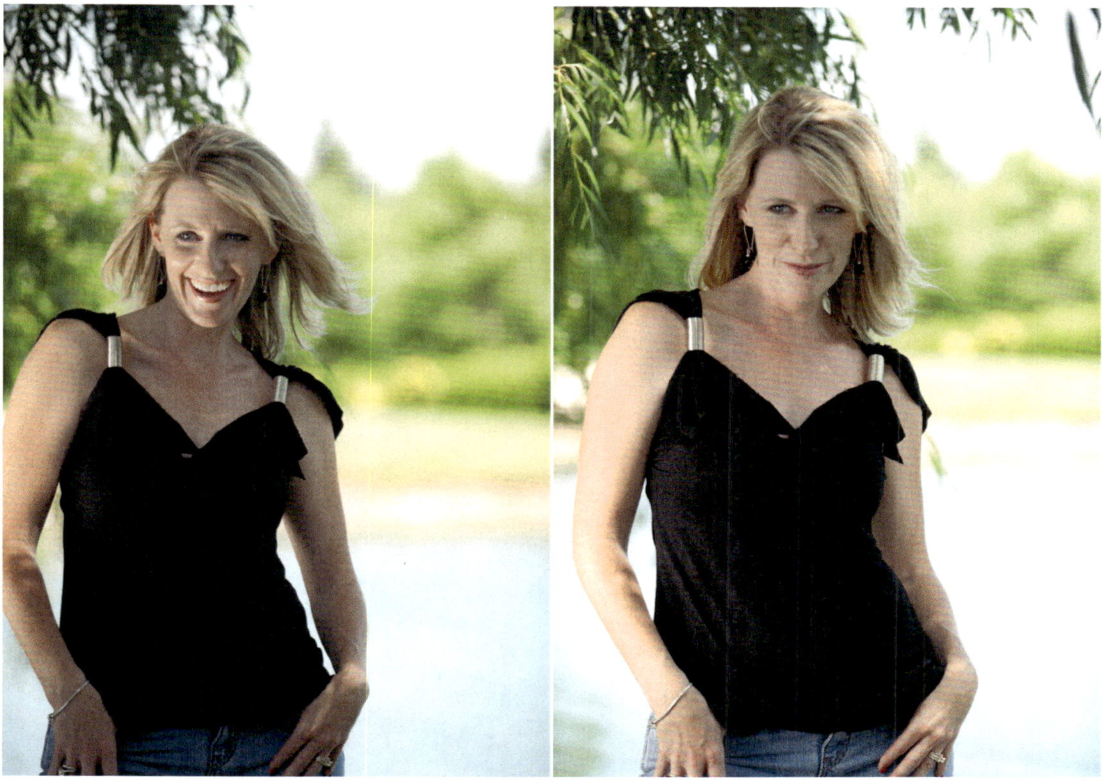

Figures 8-16 and 8-17. *Over-done lighting in 8-16, but the pose is worth it!*

Finally, here are two more shots showing a reflector set-up and the result. Note that lighting isn't all that's involved here; the curve of the pond helps frame the picture, as do the hanging willow fronds. Figure 8-19 (bottom shot) shows the result.

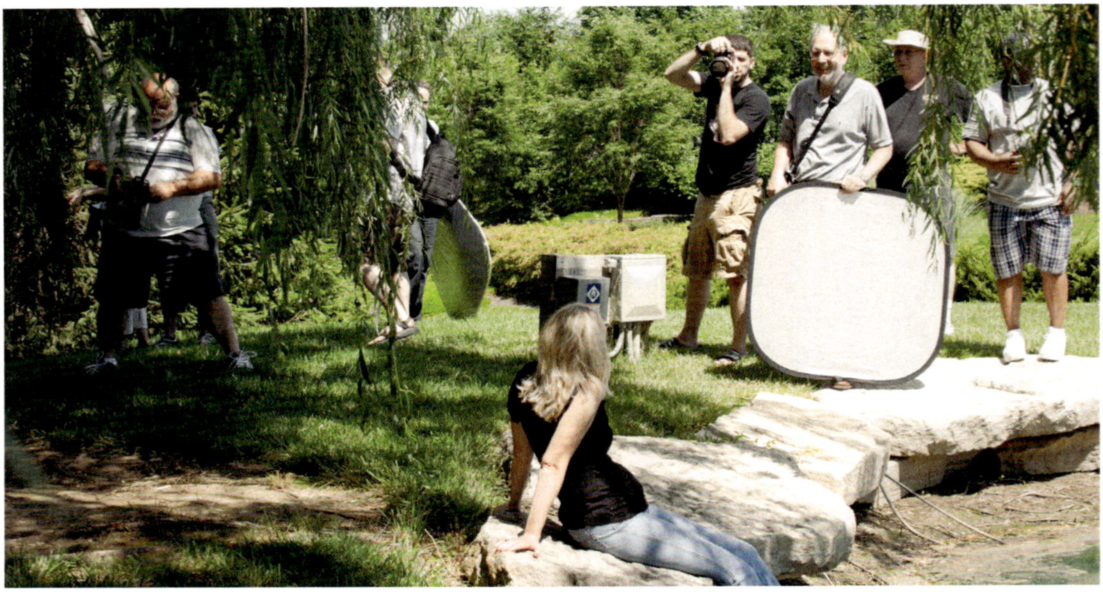

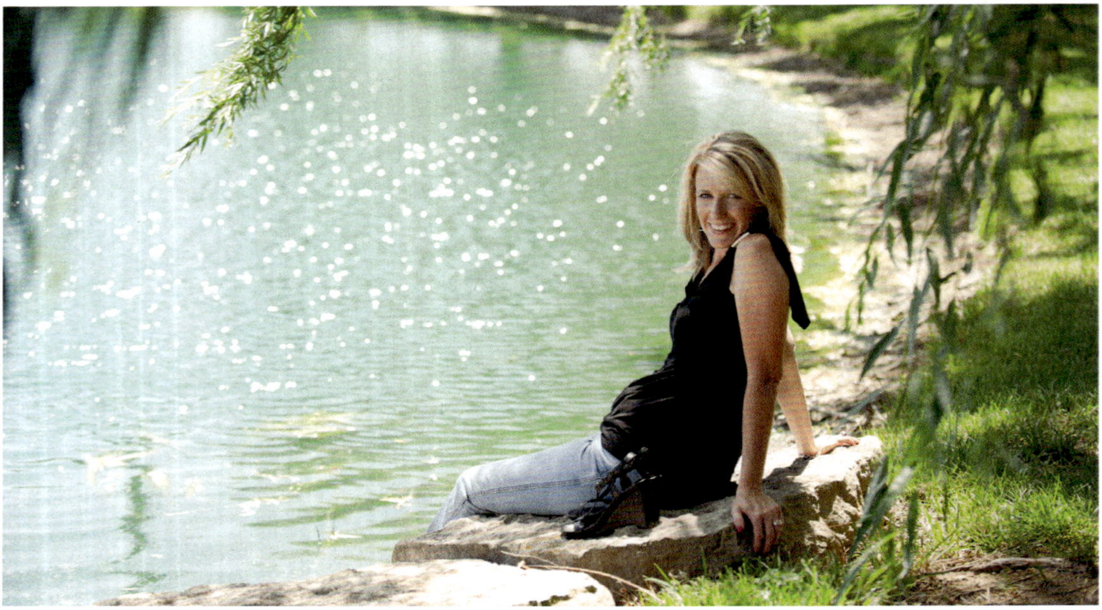

Figures 8-18 and 8-19. *8-18 is the set-up, 8-19 the result.*

Here are two shots showing the difference a little patience and a small change in position can make. Even though the client loved the first shot, it's a bad photograph compared to the second one. The sun was late, giving that golden glow, but the shadows are harsh and unflattering. The woman should have been posed as in the second photo to make her look slimmer. The second one was taken a couple of minutes later. The sun had ducked behind a hazy cloud, which still left enough light for a hint of rim lighting on the man's shoulders, but softened all those shadows. The pose is far more flattering (different woman, same guy).

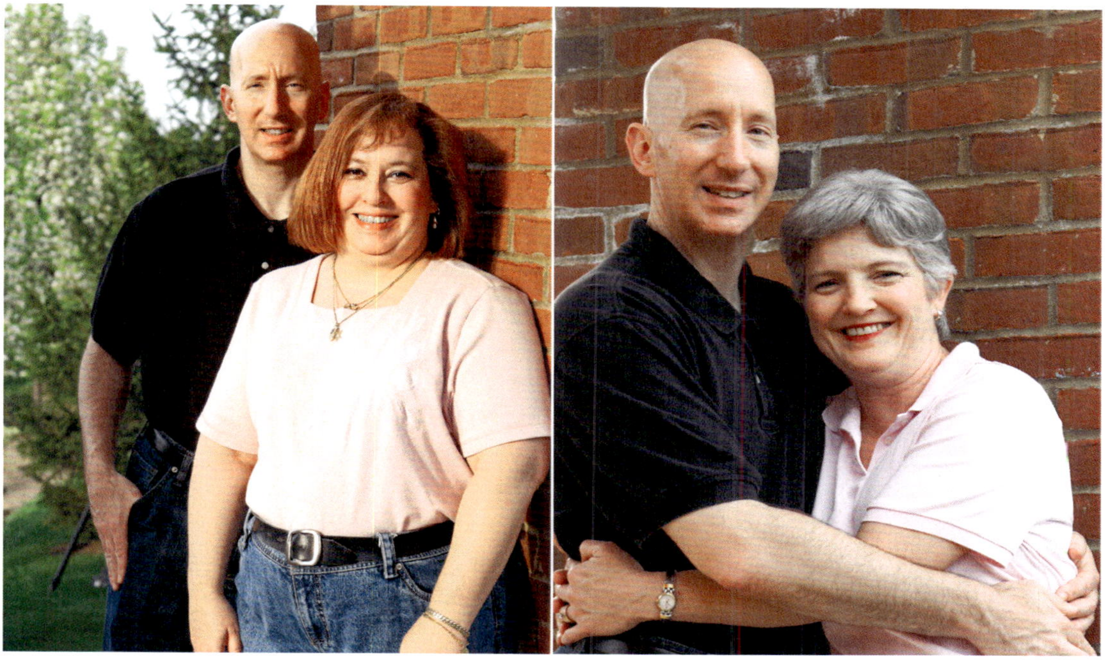

Figures 8-20 and 8-21. *High, bright overcast is better for people pix than bright sun.*

So far I've been using examples from formal shoots, simply because they illustrate the points at hand about lighting. Most outdoor people shots are far more spontaneous. Compositional elements become more important, because such shots aren't necessarily posed. Here are two good examples.

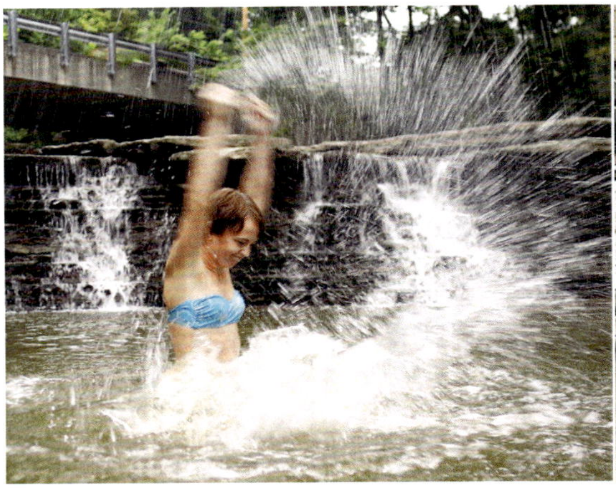

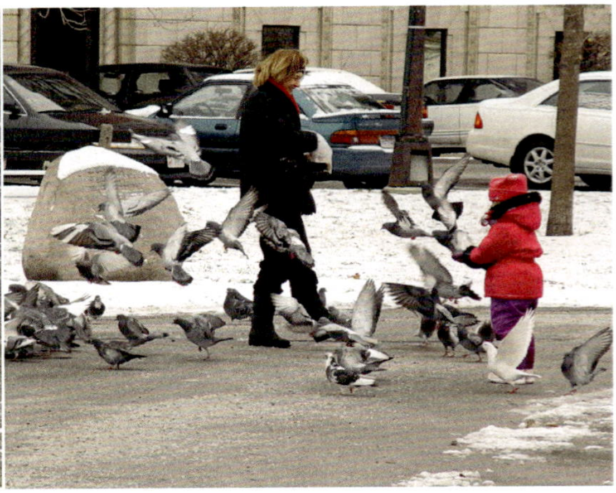

Figure 8-22. *Flat Rock, Morrow, OH. F2.8, 1/45 second, ISO 250.* ***Figure 8-23.*** *Hyde Park, Chicago. F3.5, 1/500, ISO 100250.*

Figure 8-22 was also taken in high cloudy conditions. The motion blur makes it! (f2.8, 1/45 second, ISO 250). Figure 8-23 relies on the implied diagonal between mom and daughter, as well as on the innocence and fun of the shot itself. The cloudy day helped, intensifying the color of the girl's coat.

This article could go on forever, but to sum up, key points in shooting people outdoors are

1. Manage your lighting, whether by finding the right spot, the right time of day, or the right weather conditions. If you can't do that, give nature a hand with reflectors and fill flash

2. The same compositional techniques covered in other articles ("Composition and Design" and "Story-Telling") apply in people pictures as well, and are, if anything, even more important. Perfect exposure of a bad photo gets you nowhere; you still need your eye and your sense of timing.

Just for fun, I'll end with two shots. Figure 8-24 shows kids pelting down the street when school lets out in Saluzzo, Italy. The out-thrust arm of the boy in the lower right corner says it all! Figure 8-25 uses a wide angle lens to distort the people who are gazing at their differently distorted images in Chicago's Millennium Park. Out door people pictures give the photographer room for spontaneity and creativity — at the same time!

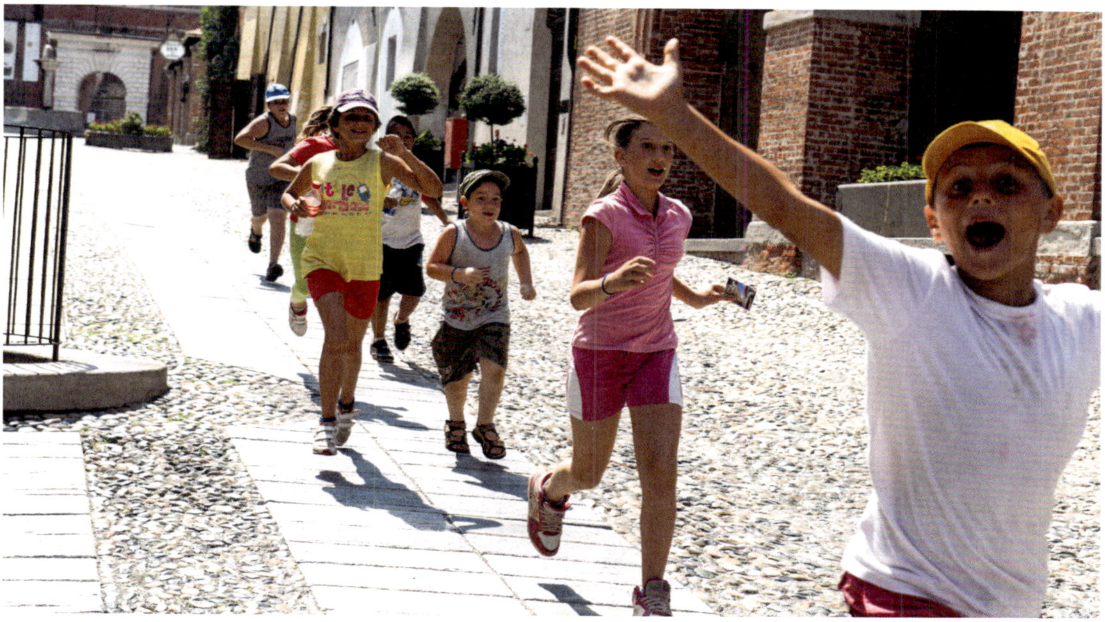

Figure 8-24. *School is out, Saluzzo Italy. F9,1/1250, 70mm, ISO 640*

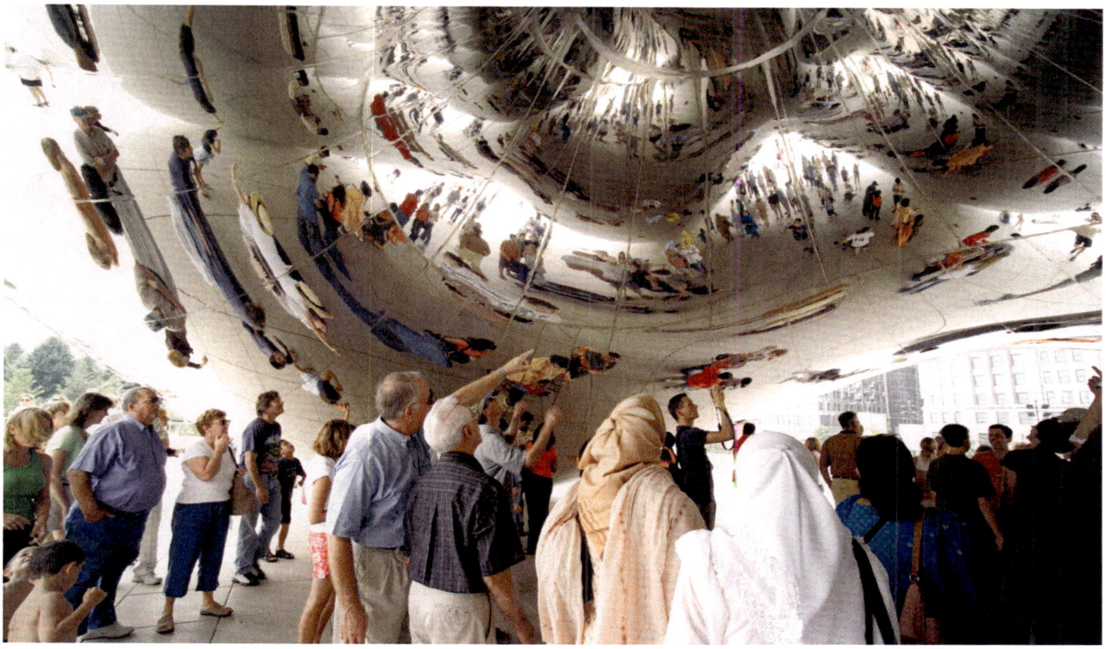

Figure 8-25. *Chicago Millennium Park, hazy sunshine. F6.7, 1/45, ISO 160. Wide angle lens creates this happy distortion.*

SHOOTING WATERFALLS

Like many photographers, I love waterfalls, and over the years I've had a lot of experience in capturing them. This article deals with some of their magic and some of the techniques for photographing them.

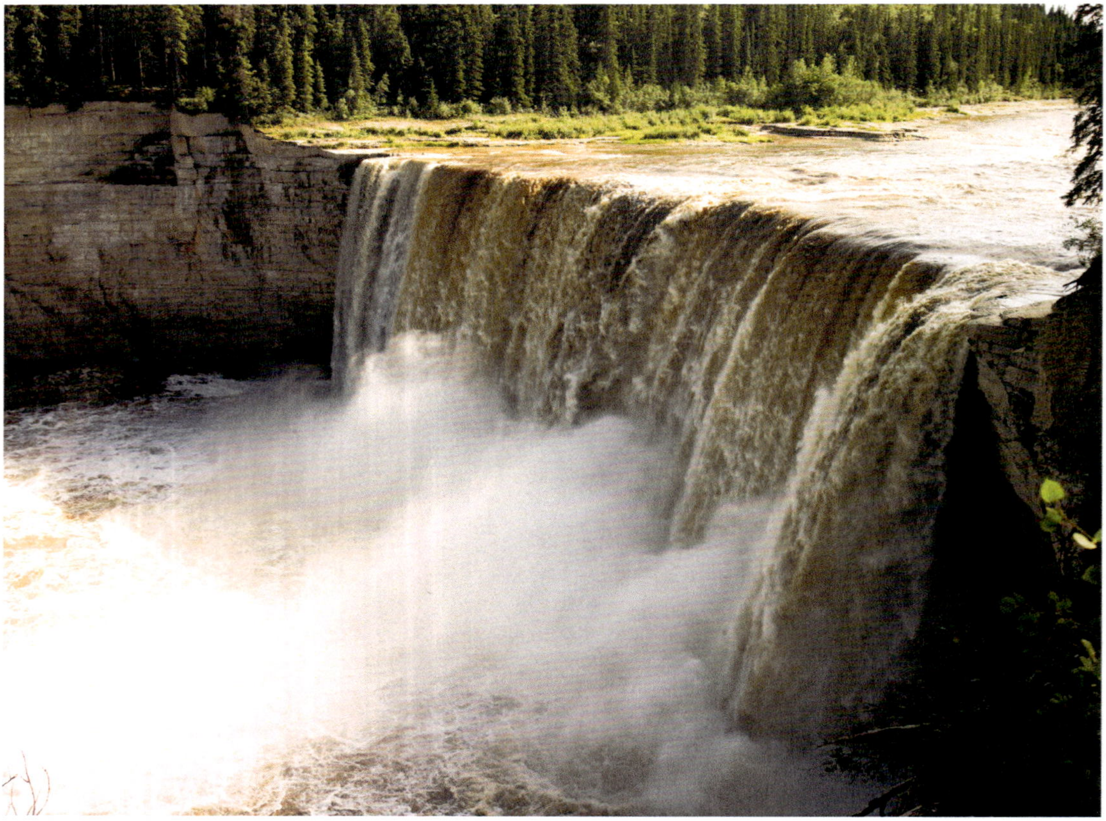

Figure 8-26. *Louise and Alexandra Falls, Northwest Territories, CA. F11,1/90, ISO 200*

Waterfalls come in a number of types: horsetails (thin vertical falls), punchbowls (where water pours from a "spout" into a bowl-shaped catchment), and cascades (step-like series of falls) are three of the most common. Broad falls, like the falls in Canada's NW territories, above, are also common.

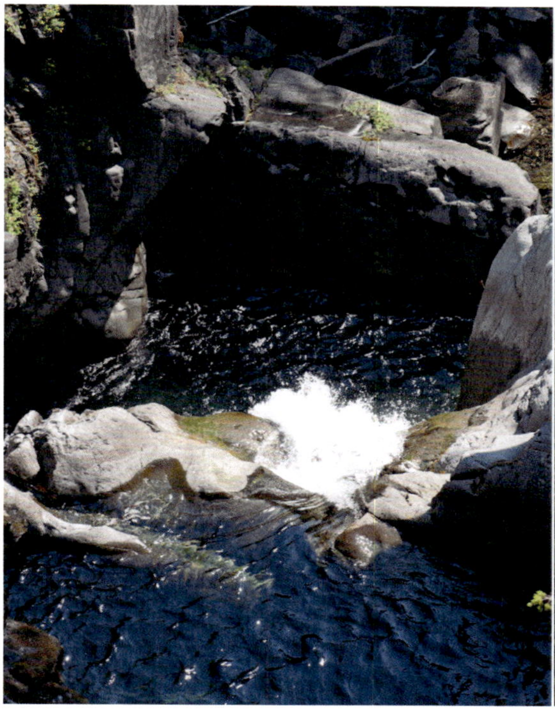

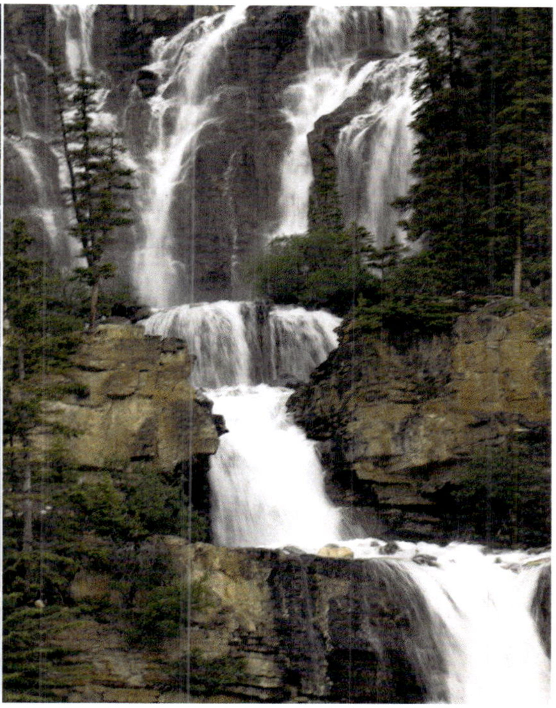

Figures 8-27 and 8-28. *Here are a punchbowl falls named Hama Hama in Washington State and a cascade in Alberta, CA.*

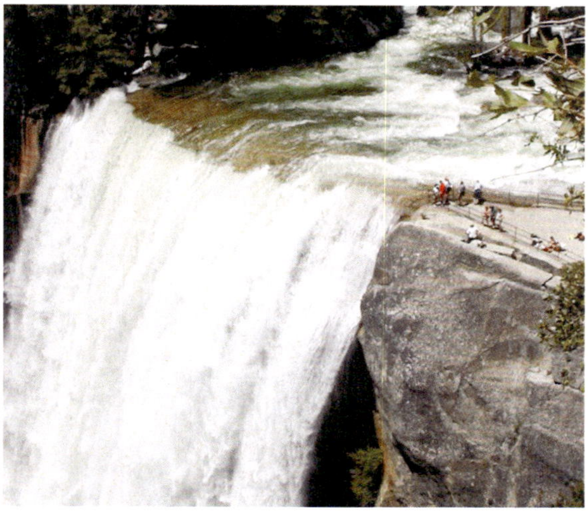

Figure 8-29. *Vernal Falls, Yosemite Nat'l Park. Note how the tiny figures give the scale.*

Waterfalls pose great challenges to the photographer. First is access. The Creator rarely put these gems where they could be seen without effort. The various park services have tried to improve on the Creator's notion of access, but the fact remains: the best ones are often the hardest to get to. But there are still plenty within easy reach of paths and parking areas.

On the general subject of access, waterfalls are dangerous places. This picture shows Vernal Falls in Yosemite National Park. In 2011 three people were killed, swept over these falls. When this picture was taken, one photographer COULD have been killed, craning his damn fool neck too far over the cliff edge!

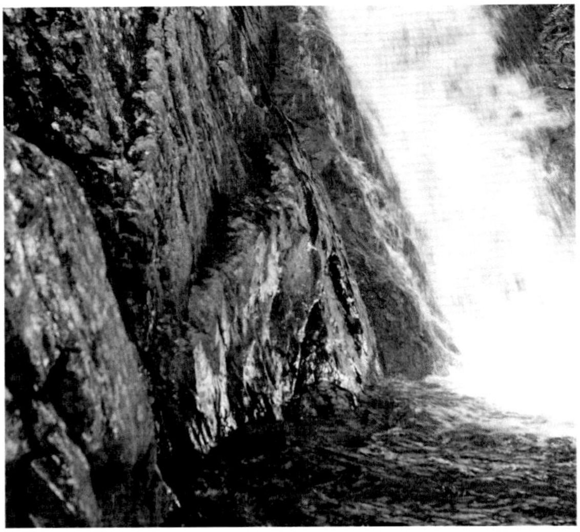

Figure 8-30. *Dunns Fall*

If you're on top of a waterfall, one slip can be fatal. If you're at the bottom, broken legs are easy to get from stumbling or tripping on the slippery, wet rocks you'll usually find. In the middle, you're often clinging to a cliff and hoping for the best. Carrying a length of stout rope is a REALLY good idea when visiting falls if you're going to be scrambling on wet rocks. If you do have a buddy, carry walkie-talkies. Cell phones may not work.

After access, the second great challenge in capturing waterfalls is achieving satisfactory lighting. By the nature of things, waterfalls are nearly white, and their surroundings typically are dark. This gives cameras and photographers heartburn. If you shoot to expose the water properly, the surroundings will be too dark. If you shoot for the overall environment, you will wind up either over or underexposed. The problem is that the range from light to dark in most waterfall scenes is greater than the camera can handle. Our eyes can manage, but few cameras can handle this sort of range.

The third big challenge is getting a decent, unobstructed view while still staying safe and dry (or dry-ish). Waterfalls make a great place for view-obstructing logs to lodge. They also encourage spindly bushes with far too many leaves to grow where they have no business being! Sometimes the problem of obstructions and access can force you into a place with limited visibility. A small folding brush saw is an excellent idea, but cut only what you must to get a sight line! Be discreet, and damage not the environment.

So, here are some tips for dealing with the various challenges.

First, decide what it you want to convey about the waterfall: its shape? Its romantic quality? Its power?

These two pictures illustrate the different "feel" you can get. Both shots are of a small waterfall in Brown County, Ohio. The chief difference between them is shutter speed.... but the feeling of each shot is completely different.

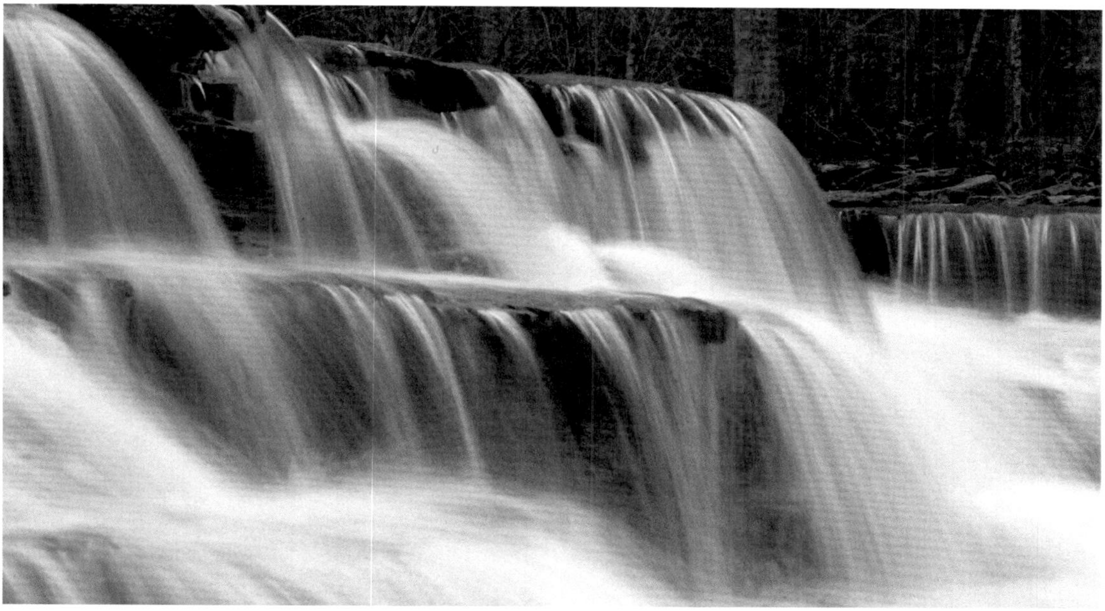

Figure 8-31. *Small un-named falls, Brown County OH. Exposure: f 32 +.3, 1 second, ISO 50*

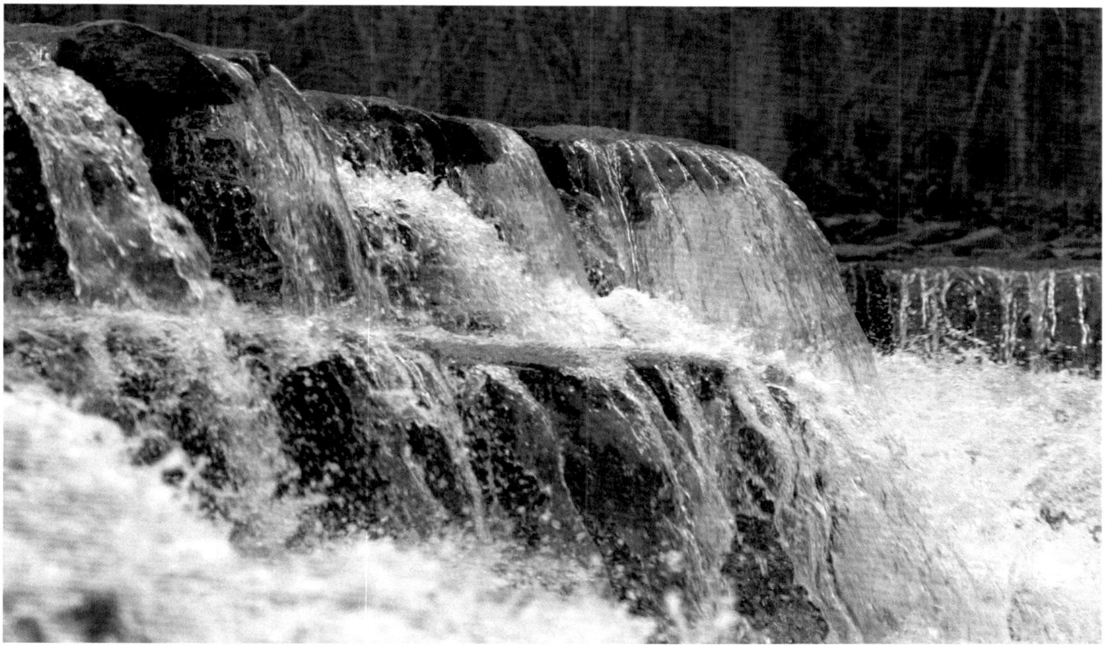

Figure 8-32. *Same as Figure 28, except exposure is changed to F 9.5, +/67,1/250, ISO 200*

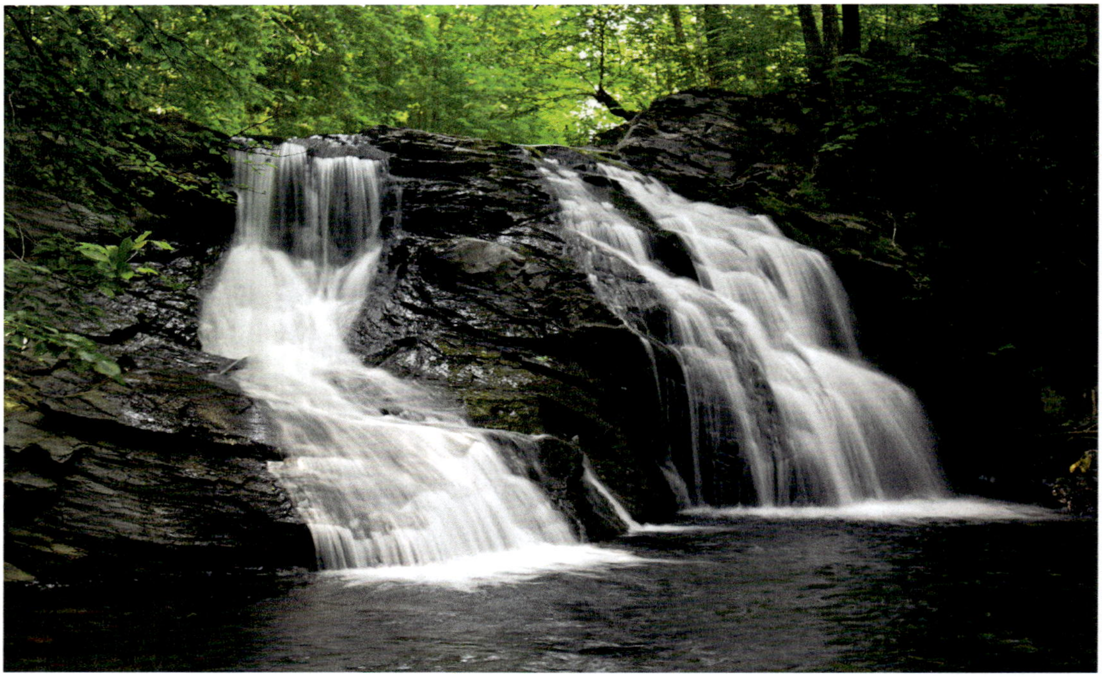

Figure 8-33. *Putney Falls, Putney, VT. Part of a truly lovely series of cascades. F6.3,+.3, 1/3 second, ISO 100*

If the waterfall is a cascade or plunge, or is in a very romantic setting, you may want to get that smooth and foamy look, as in the Putney, VT falls above. To get this look you need long exposures. It's best to shoot in cloudy conditions, or just before dawn or at dusk. Bright sunlight is your enemy. You can use ND (neutral density) filters to darken the scene to give you a slower shutter speed; polarizers can also help in this regard. In the shot above, the tree cover helped limit the sunlight. A polarizer and an ND#9 filter were used. The exposure was ISO 100, f6.8, 1/3 second.

To get the foamy look, you will need to use a tripod and an exposure time of about 1/3 to 1/2 of a second, or even longer. To get shutter speeds this slow, shoot when it is cloudy or very early or late in the day. Use a neutral density filter, or two polarizing filters stacked together at right angles, and your slowest ISO setting. Most often you will shoot at the highest aperture you've got, f22 or f28 or f32, especially if you've got sun to worry about. What you can do is set the exposure, then compensate up by up by 1 to 2 f-stops. This will let the whites be whiter without totally killing the darker background.... usually.

Again, you cannot get good results for the "foamy" look without a tripod. And, while you're at it, use time delay and mirror pre-up if you have it to minimize vibration. And don't shoot for foamy on a windy day; all the vegetation will look blurred, as will the falls themselves.

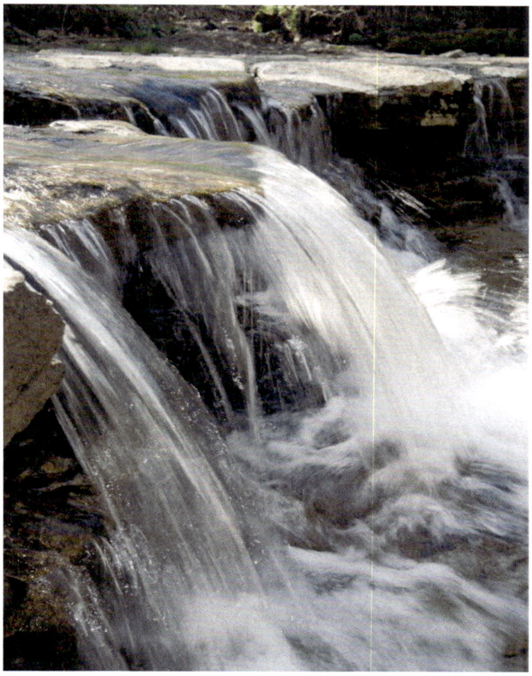

Figure 8-34. *Bear Creek, Foster, OH. Flash placed behind water, lower left out of frame. F10,1/25, ISO 100*

It's a serious waste to try and get a "foamy" picture of a very large falls any distance away. That's because distance already smoothes out the water for you! You are now free to shoot whatever exposure you like. I've found that shutter speeds between 1/80 and 1/250th of a second are good for capturing the whole scene or distant falls; adjust your aperture appropriately.

Believe it or not, you can even use flash in shooting waterfalls! Put a flash inside two or three Ziploc bags, along with a radio trigger (I use Pocket Wizard), and stick it behind a curtain of water or a rock. This trick works wonders, but wet Pocket Wizards cost a ton to replace, so be careful. Here's an example: See the luminous quality to the water and rocks in the lower left? That's from a fill flash strategically placed.

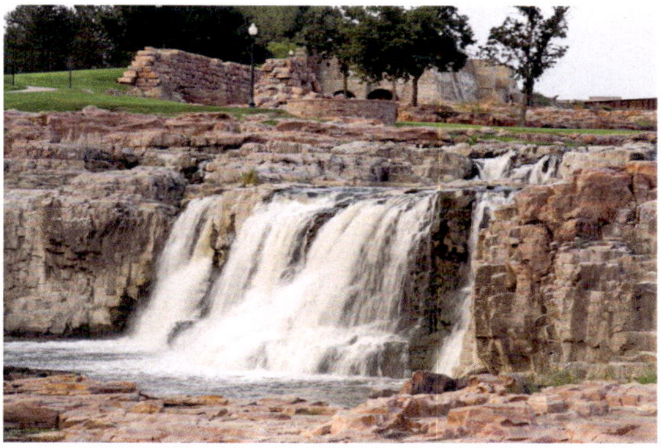

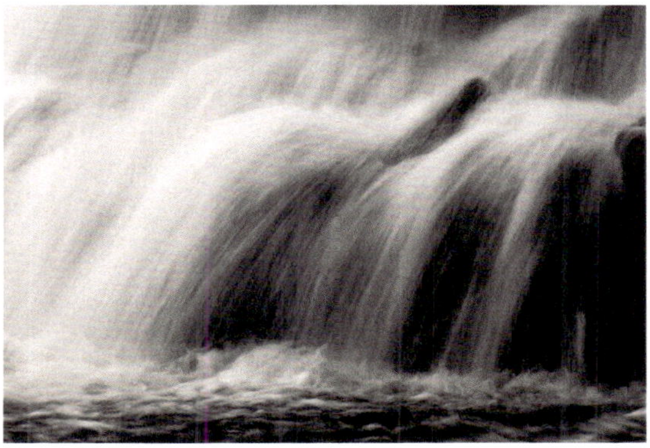

Figures 8-35 and 8-26. *Sioux Falls, Iowa. Wide view (8-35) shot at f 13,1/20,ISO 800. Close-up (8-36) shot at f14,1/8, ISO 125. Tripod used in both shots.*

If you want to capture the power of the fall, or if view is obstructed, try focusing only on a particular part of the fall, like the Vernal Falls picture, Figure 8-29. You can get some amazing results this way! Here's an example, Sioux Falls in Iowa. The whole view is nice, but the close-up has a wholly different feel.

Below, Figure 8-37: Sometimes, less is more. Figure 8-36 is graceful, almost wistful; this one is about power. This is the top half of a falls up a trail along Mineral Creek, outside Valdez, AK. Exposure was f5.6, ISO 1000, 1/400 second. The idea was to capture the energy of the water plummeting over the basalt blocks making up the falls.

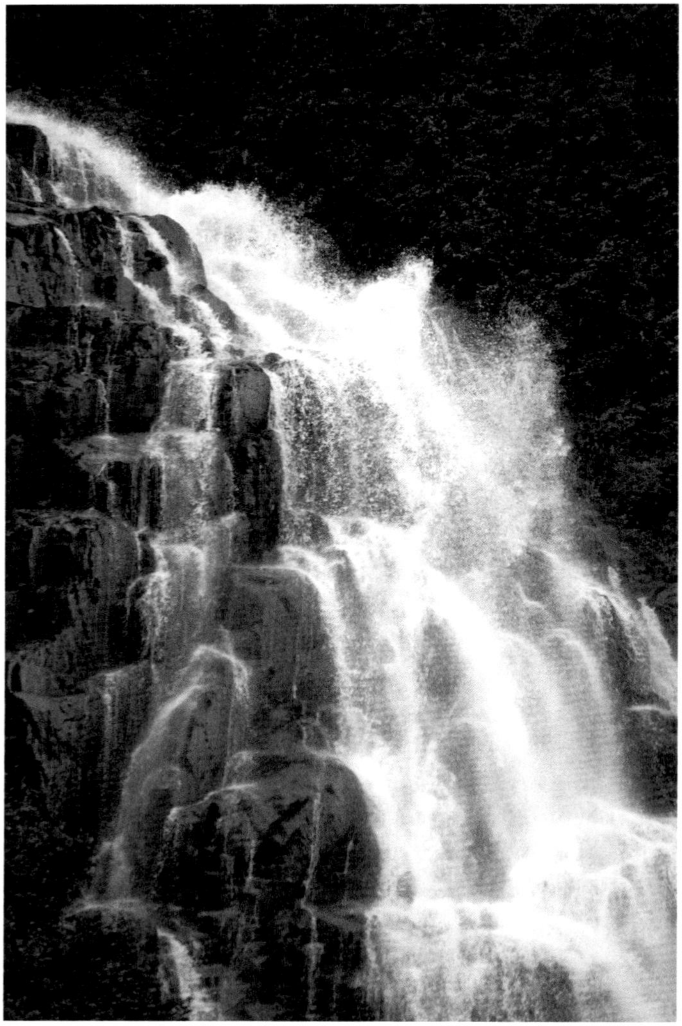

Figure 8-37. Mineral Creek near Valdez, AK

Sometimes you really do want the whole thing. I can't leave this article without this shot of a nameless falls plummeting about 600 feet to the sea in Molokai, HI. Note the self-foaming quality of the water. Exposure is f4.0 at 1/600 of a second. It's the distance, not the exposure, that makes the water foamy in this case – but I needed the shutter speed to offset vibration in the chopper from which it was shot.

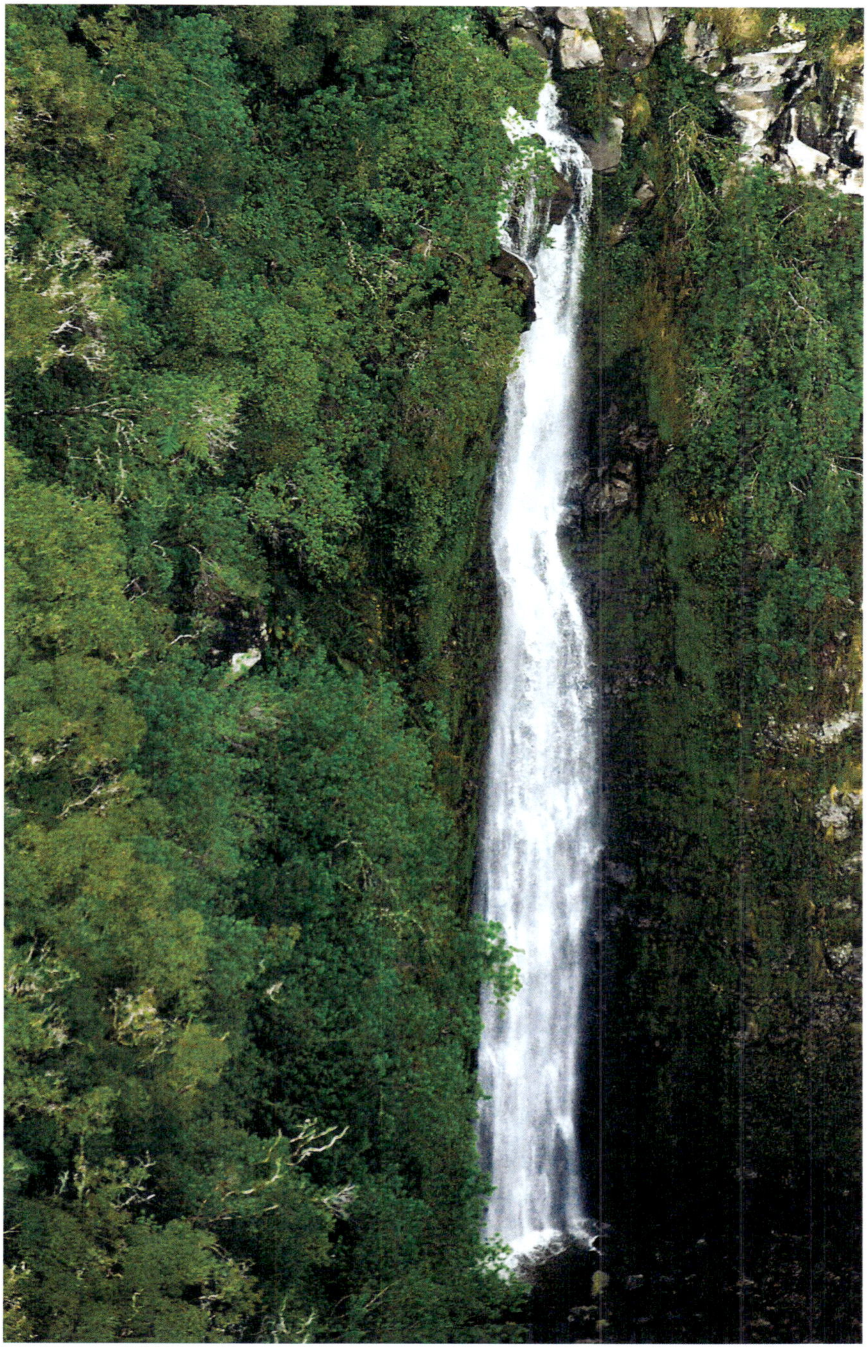

Figure 8-38. *Nameless horsetail falls on Molokai, HI. Distance foams the water. F4, 1/600, ISO 200.*

CHAPTER 9: *Editing and Workflow*

GOOD PRACTICES FOR DOWNLOADING AND STORING PHOTOS
TOO MUCH OF A GOOD THING? (CULLING AND GRADING YOUR PICTURES)
CROP YOUR WAY TO BETTER PHOTOS
CONCEPTS IN PHOTO EDITING
HDR (HIGHLY DUBIOUS RETOUCHING)
CASE HISTORY OF A FINE ART LANDSCAPE

CHAPTER 9: *Editing and Workflow*

GOOD PRACTICES FOR DOWNLOADING AND STORING IMAGES

The most horrible sound you can hear is "splish." That's the sound of your memory card slipping out of your fingers and into the murky waters of the pond.

Bad enough to lose what you've just shot, but a nightmare to lose the last two days' worth because you didn't download and back 'em up last night.

Memory cards are amazingly durable, especially the CF cards. I've had them run through the laundry and work fine. They've flopped around loose in my pocket with wild-growing lint all around them. I've even had one fall into a cat box. In all three instances, they kept on working. But they're not bulletproof, they can be shorted out, cracked, pulverized, melted, and crusted in engine grease.

All right, all right, you're probably exclaiming. Get on with it, already.

To avoid losing my images, I download them to a hard drive the minute I get done for the day, whether at home or on the road. Then I check to make sure they are really there. Later, they're copied to a second external hard drive.

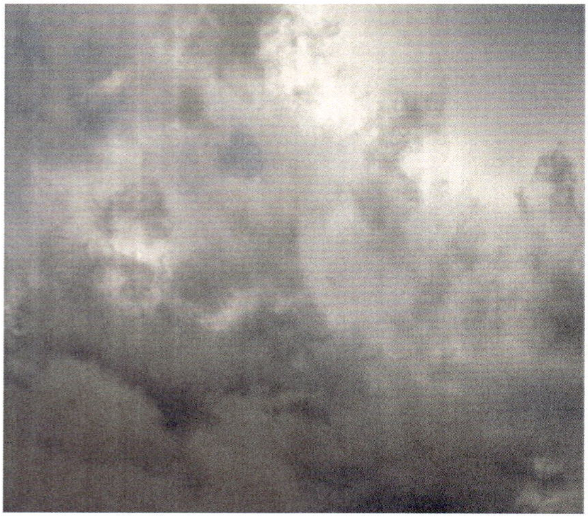

Figure 9-1. *The cloud*

Ideally, you would keep the two storage locations in separate places. If there's a fire, you can easily lose both your computer with its internal drive and an external drive. Better to keep them in separate buildings, though few of us actually do so. Or you could try cloud storage — storing your images on someone's server for a (generally modest) monthly or annual fee. It looks like this: you are in the lower left, down in the accident-prone gloom, and your photos are somewhere vague in the lighter part of the picture.

I use cloud storage myself myself for secondary not primary, storage, and it seems to be the coming thing. Cloud storage has security risks;

anything can be hacked. It has longevity risks: if your supplier goes out of business, where are your pictures? If you don't pay your bill, how long do you have before your account is wiped clean?

Cloud storage also takes an ungodly amount of time to load to unless you have a full business class ISP agreement. If you shoot large pictures in RAW, as I do, you can easily spend hours and hours uploading a day's work. Not a good use of time. Finally, it presumes you have a web connection. If you don't, you can't access your property.

On the plus side, cloud storage does get your images out of the house, so you do have duplication that is physically pretty safe.

One such storage / image sharing site is Smugmug.com. If you're a Smugmug user you can get something called "Smugmug Vault" for a very reasonable cost. Even without the Smugmug Vault service, your pictures stay on their servers forever and you can download them at original size straight to your computer. This has been a really useful feature from time to time, including sourcing pictures for this book. But I don't load everything up there, just the stuff that's for public consumption.

The solution I use while on the road is a portable external hard drive. Then when I get home, it's into the computer and from there to a Drobo. A Drobo (from Data Robotics, hence the name) is a black box with slots for 4 or more internal hard drives which the machine treats as an array. Lose one drive, no problem. Yank it out, slide in another one, and wait while the Drobo rebuilds the array. That can take many hours, but you can use your computer (and the Drobo) in the meantime.

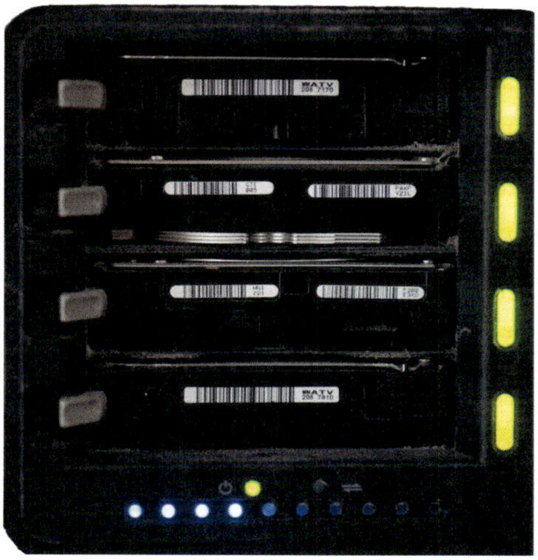

Figure 9-2. *Drobo with cover removed to show 4 1-terabyte drives. Lights on bottom show how much capacity has been used. This unit is 40% full.*

I use four, one-terabyte drives in my Drobo. That seems to be plenty, but I can always expand.

What I don't do is store anything but photos there. I use a different internal hard drive to back up my OS and data files. That's faster and more secure, as well as keeping my Drobo from clogging up with multiple iterations of the same thing.

Saving an image is no good unless you can find it later. It's all well and good to have strong discipline about downloading. You need similar discipline in designing the filing system you want to use to keep your stuff where you can find it easily.

This is not an area where I am particularly qualified to give good advice, since I have trouble following it myself.

But in general, I set up files into two broad categories: clients and non-clients. Then annually I create new folders for the year. Within the Client folder, sub-folders are given the client's name, a one-word job description, and the date. On the non-client side, it's by job, event, trip, location, or subject along with a date, e.g., Non-client 2009, Washington State, Hama-Hama.

That way, if I can remember who, or where, or when a picture was taken, I can find it. For the keepers I go further and keyword them. Makes it easy to find by subject.

I am a Mac user, but I don't use iPhoto. iPhoto stores a complete copy of your photo, not a reference copy or alias, so it takes up massive amounts of hard drive space. Worse, you have to work with Apple's scheme for finding things ... not a bad scheme, I just don't like it. Thousands do, and if it works for them (or you), God bless.

The point is to get the good stuff out of your camera and into safety as soon as you can. Then sort out the treasures from the trash (see "Too Much of a Good Thing?" elsewhere in this book) and duplicate the treasures to physically safe storage.

TOO MUCH OF A GOOD THING? (CULLING AND GRADING YOUR PICTURES)

When I get back from my travels, I face the challenge of culling the photos. This problem of winnowing out the best from the merely very good is one we all face. In my case, I had it pounded into me at an impressionable age to NEVER THROW AWAY ORGINAL DATA. That conditioning makes it very hard for me to ruthlessly toss stuff, and is one reason I shoot in RAW 90% of the time. But I try, really I do.

In the age of digital photography, you'd better throw out about 90% or invest heavily in terabyte size storage. So assuming you're going to sort it all out, how do you do it? I sort for technical competence (no blurry pix live on in my computer). I sort for story-telling, impact, and personal relevance, in that order. All of this is actually very holistic.

What I actually do is use Adobe Bridge (but there are lots of other choices) to run a slide show of that day's work. Then I assign 0, 3, 4, or 5 stars to each image. Once I've done that, I throw out the no-stars and 3-stars. I look again at the 5-star pix, and this time give them color-coded labels based on quality—red, yellow, or green. The greens are the top of the heap.

This screen shot should make it pretty plain.

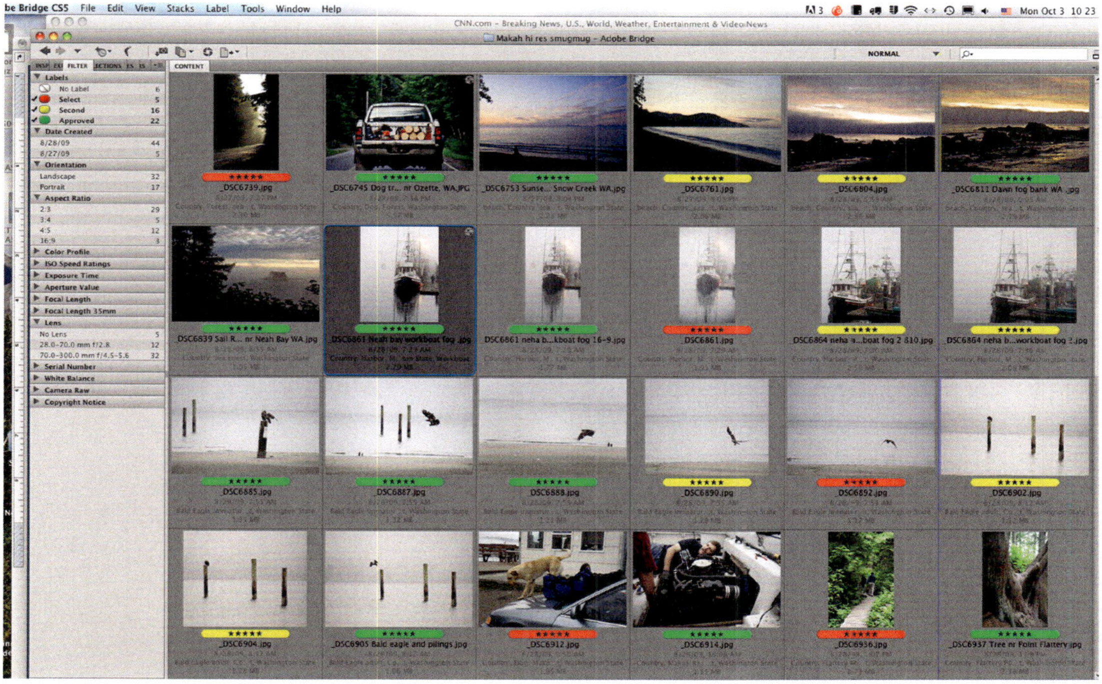

***Figure 9-3.** Screen shot showing one method of identifying keepers.*

At this point if there's an image that's questionable, or that I think could readily be improved, I look at it in Camera Raw, poke it a bit, and if I like it, it gets the coveted colored label.

If this is a portrait shoot, or an event, I stop here. If it's my own artwork, I put the whole thing away for a day or two and visit the pix again. I also review the 4-stars, seeing if any should move up or out.

The time lapse is important in editing prose or photos. Things can look very different when a little cool air gets between you and your fevered ego.

The final step is to identify the very best, then edit them and run test prints if necessary.

Now here are a few numbers. When I started shooting seriously, I became a dedicated overshooter. I took way too many duplicated pictures, and I do mean duplicated. That's not the same thing as re-framing, moving a step, or bracketing exposures. It's what they call "chimping" --- after the theory that an infinity of chimps with typewriters can eventually type out "Hamlet" to perfection. (Actually they probably get as far as "angels sing thee to thy djkfj8upewrpoipnanfpou8p".)

When I first started digital shooting I was lucky to get a 3% keeper rate. Now it's 10-12%, but I don't shoot as many frames unless I'm trying to solve some technical problem or other. But I wonder if that's too high, and I'm just going soft in my old age.

When photos are very good, each one may vary in small details from the others, but wind up telling a different or better story. For me, it becomes harder and harder to choose that best one or two from 8 really good ones. In fact, it becomes a more subjective process. But I lack confidence ... don't want to misjudge and risk throwing out a baby when I meant to toss the bathwater. So I tend to keep too many of the merely good.

The important thing is to make sure that the merely good ARE good and absolutely to pitch anything that isn't. The star-and-label system seems to work pretty well for this.

For more details on photo editing, see "Concepts of Photo Editing" elsewhere in this chapter.

CROP YOUR WAY TO BETTER PHOTOS

Cropping a photo is often the best way to improve it. But cropping, useful as it is, also presents some problems. This article deals with both sides of the cropping issue.

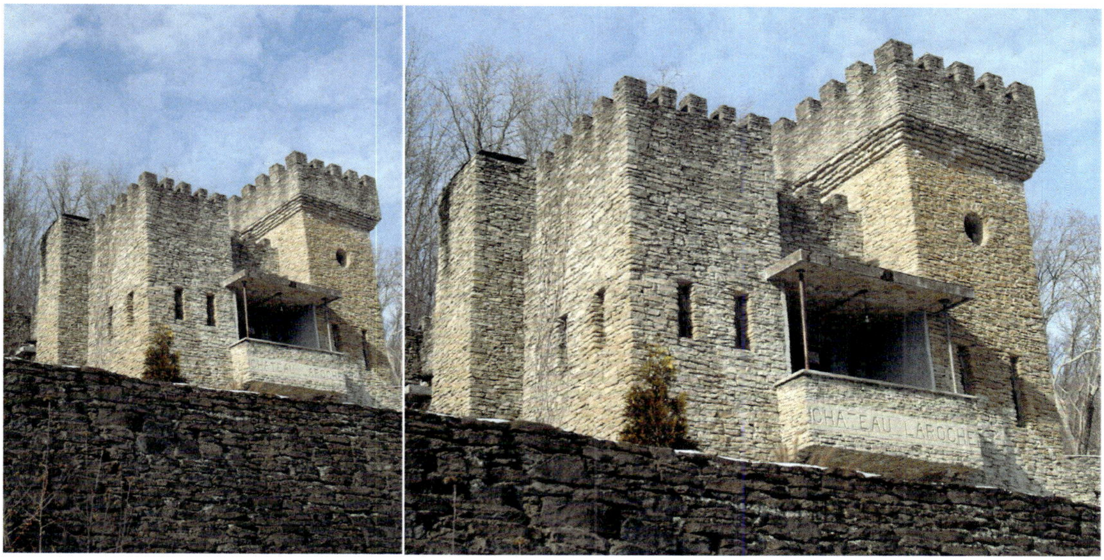

Figures 9-4 and 9-5. *Loveland Castle, Loveland OH. 95 is the cropped version. What a difference!*

First, in case you've forgotten, cropping a photo is the art of removing parts of the image, leaving a residue which is necessarily smaller than the original. "Smaller" here refers to file size – the fact is that when you crop you toss out data, and that can come back to bite you if you don't follow a good workflow.

Even if there's a potential penalty to pay in cropping, there are plenty of benefits. Here's what cropping can do for you and your pictures:

- Exclude unwanted material that adds no value to the shot
- Focus on the essential elements of the picture
- Improve composition by putting the center of interest in the "right" spot
- Eliminate distractions
- Alter the story your picture is telling (and every picture tells, or should tell, a story)
- Fit your image to an appropriate size for framing, yearbook use, etc.
- Save work "fixing" portions of the picture by removing the bad areas altogether

Cropping and composition are tightly linked. It's often said that the best place to crop is in your viewfinder, prior to clicking the shutter. That's usually true, but there's a downside: if your picture is perfectly framed and perfectly composed in the camera, you cannot later reformat from, say, 3:2 to 5:4 aspect ratio without doing grievous bodily harm to your art. You'll save work (and definitely become a better photographer by perfectly composing your work in camera) but you do create some limitations by framing your picture so that there's no room to crop!

Most of us don't have that problem. Our pictures are rarely shot with perfect composition, and most of us waste much or most of the frame space with stuff that doesn't help the picture. Zoom lenses help a bit here, but primes (lenses with a single fixed focal length) are pretty well guaranteed to need cropping because you can't always move to the right spot to get the picture perfectly composed. Landscape shooters may have no way to get a shrub out of frame, for instance (though I always carry a pocket saw to help with that problem).

Let's talk about excluding unwanted material, the first reason for cropping a photo.

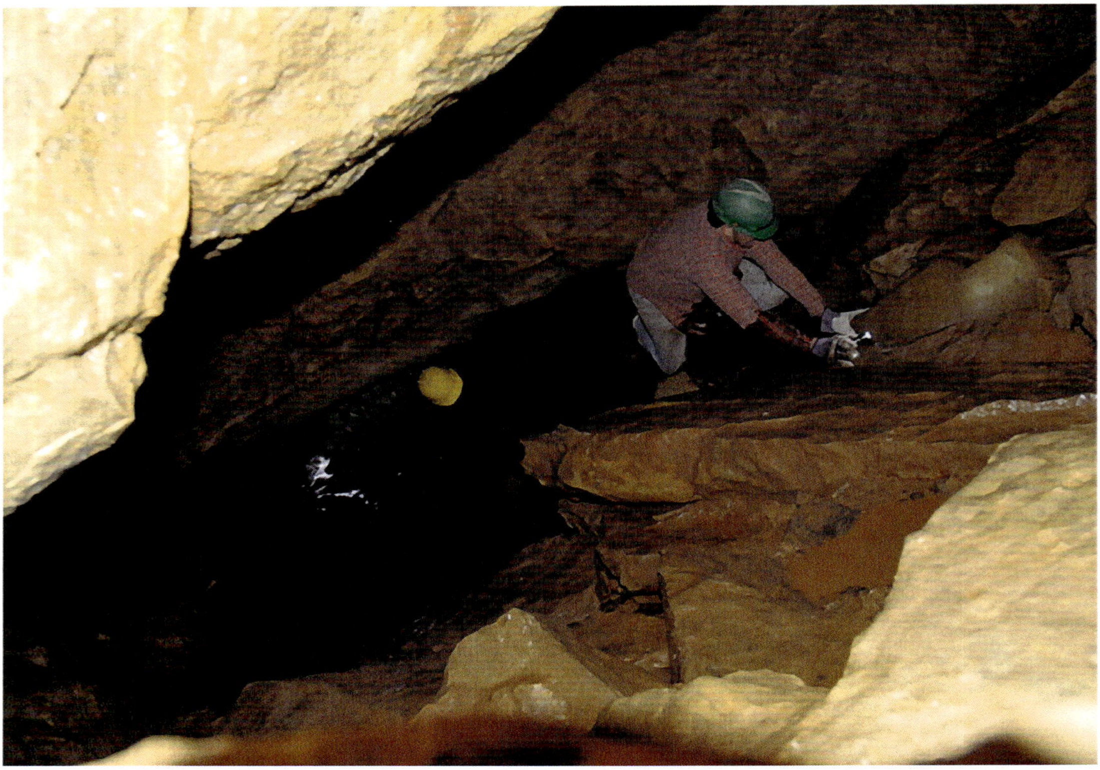

Figure 9-6. *Center of interest (the explorers) is trivialized by the very bright rocks lit up by the flash.*

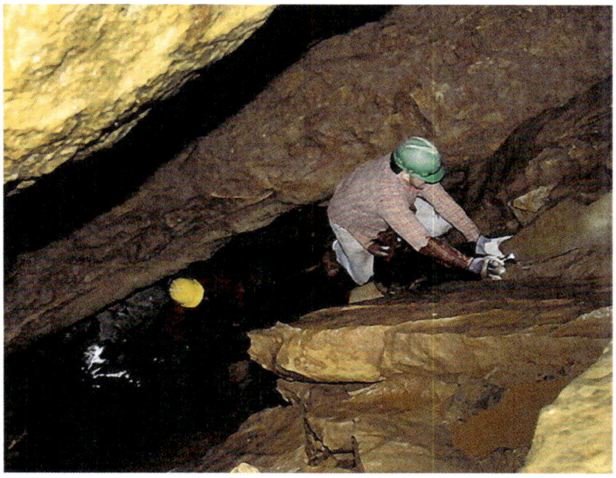

Figure 9-7. *The distracting bright rocks have been cropped out, and the central area has been brightened, while the upper left and lower right corner have been burned (darkened) to re-balance the image.*

In this cave picture used by permission of the maker, Dave Braughler, the shooter had a terrible problem – pitch-dark location, bright rocks in the foreground, nasty hard-to-light shadows in the center obscuring the central figures.

The image was cropped so as to eliminate most of the distracting bright rocks. The crop was moved to place the central subjects in the best position. Then the remaining bright rocks were burned down and the central area dodged (lightened) to increase the visual importance of the cavers.

In this case the crop achieved three of the main reasons for cropping: removing needless material, eliminating distractions, and improving the picture's story-telling. Three for one, this shot only. Here's another example:

The first shot shows the entire frame. There's more environment than bunny. The second is cropped much more tightly. This shot still shows the grasses and dirt for context, but concentrates on the bunny. Cropping the shot increases the importance of the center-of-interest and enhances the story-telling.

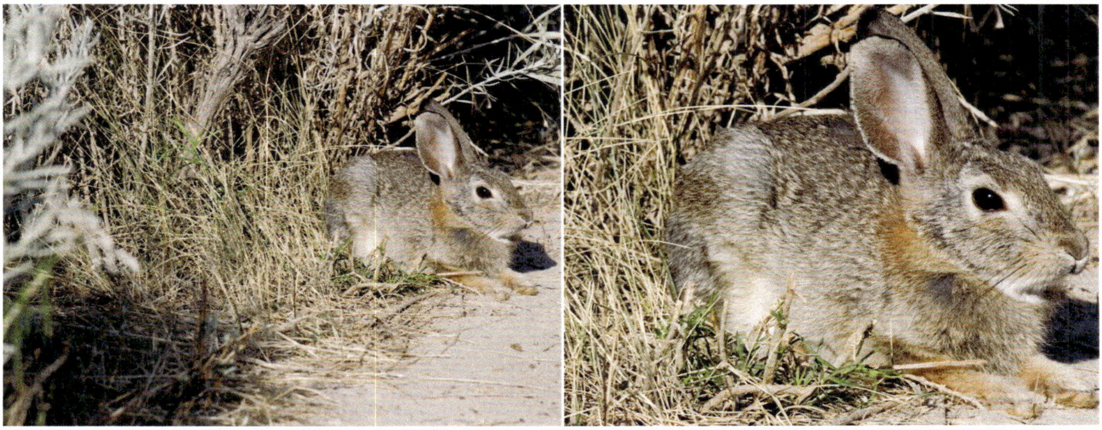

Figures 9-8 and 9-9. *The cropped version (9-9) clearly shows the dramatic difference cropping can make.*

You have to make sure the material you're cropping out is genuinely distracting, wasteful, or technically bad (like the cave rocks above). You can overdo spotlighting the central subject, as in this portrait of a girl and her cello.

In this example, the overall shot is out of balance and full of distractions. The second version eliminates the cello, and we're left with an inexplicable expression on the girl's face. If there's a story in this version, it's very dark and obscure. The last version concentrates on the girl and her cello; it's a story-telling shot showing how she and the instrument are emotionally connected.

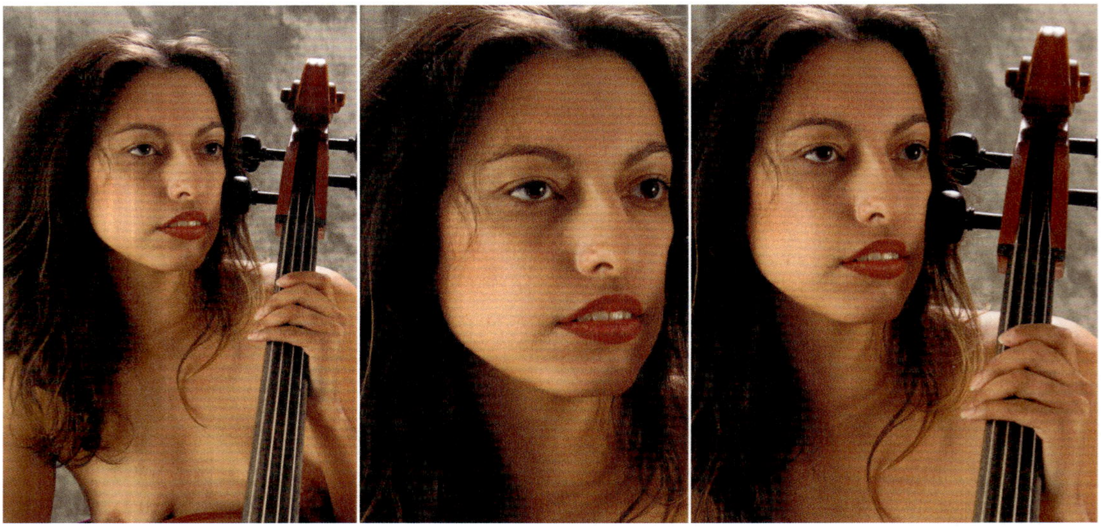

Figures 9-10, 9-11 and 9-12. *Figure 9-10 includes too much of the girl, distracting the viewer from her face. Figure 9-11 is cropped too tightly. Figure 9-12 is the "Goldilocks" version.*

Cropping for Impact

Figure 9-13 is a shot of a church window in Chimayo, New Mexico. Aside from being a little crooked, the shot lacks impact because too much of the material is irrelevant. Clearing away most of the wall surrounding the window increased both the impact and the story-telling value of the shot. It also improved the composition.

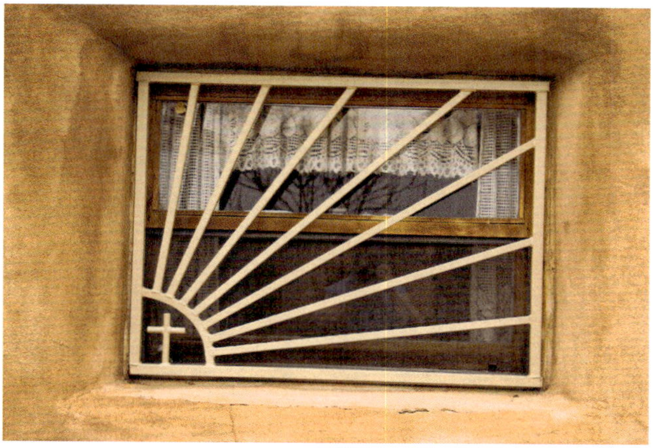

Figure 9-13. *Original*

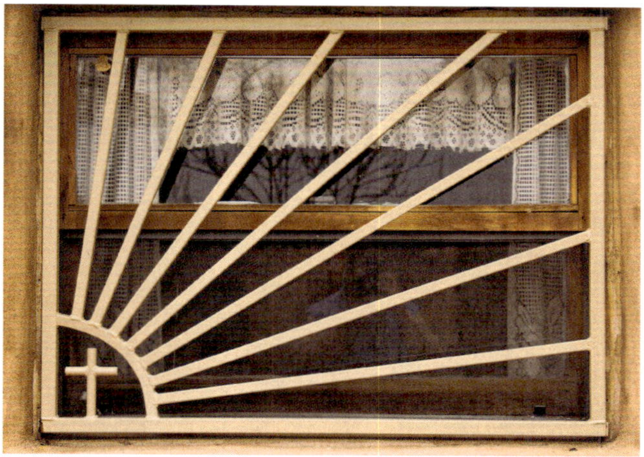

Figures 9-14. *Cropped*

Telling the right story, or telling your story right?

Sometimes you crop to enhance or exaggerate some element of your shot. In Figures 9-15 and 9-16 below, I cropped a L-O-N-G story to make a shorter one! I wanted to change the feel of the shot. Not coincidentally this shot can be used across a double-page magazine spread with room for an article underneath and for title display text in the dark areas over the horses.

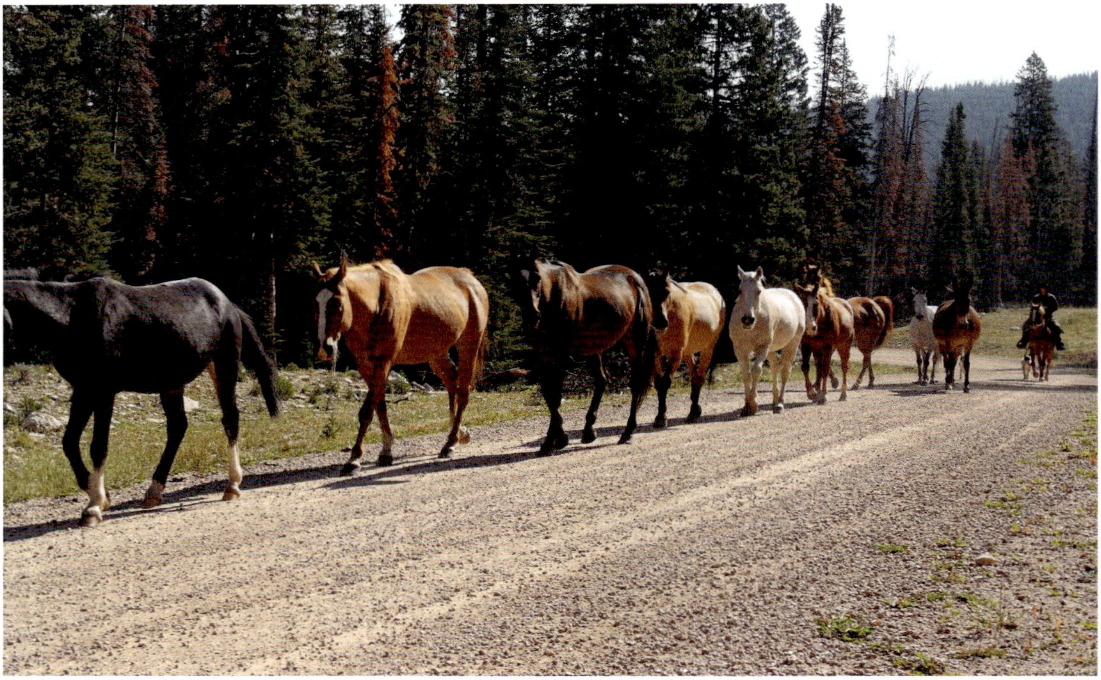

Figure 9-15. *Cowboys moving horses, Burgess Junction, WY*

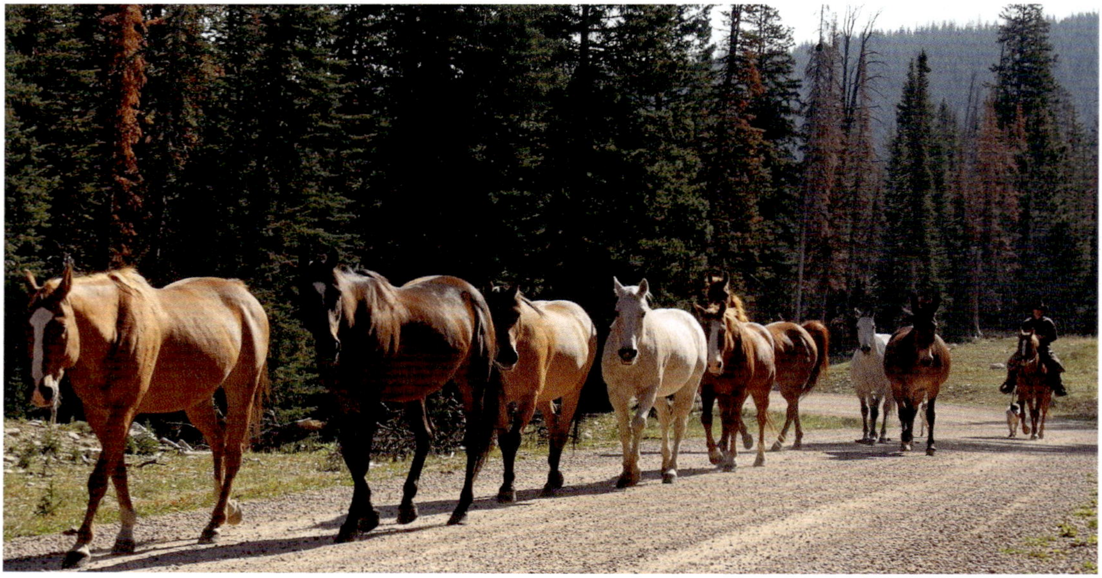

Figure 9-16. *Cropping makes the photo more of a "story" picture and provides makes it suitable for magazine use.*

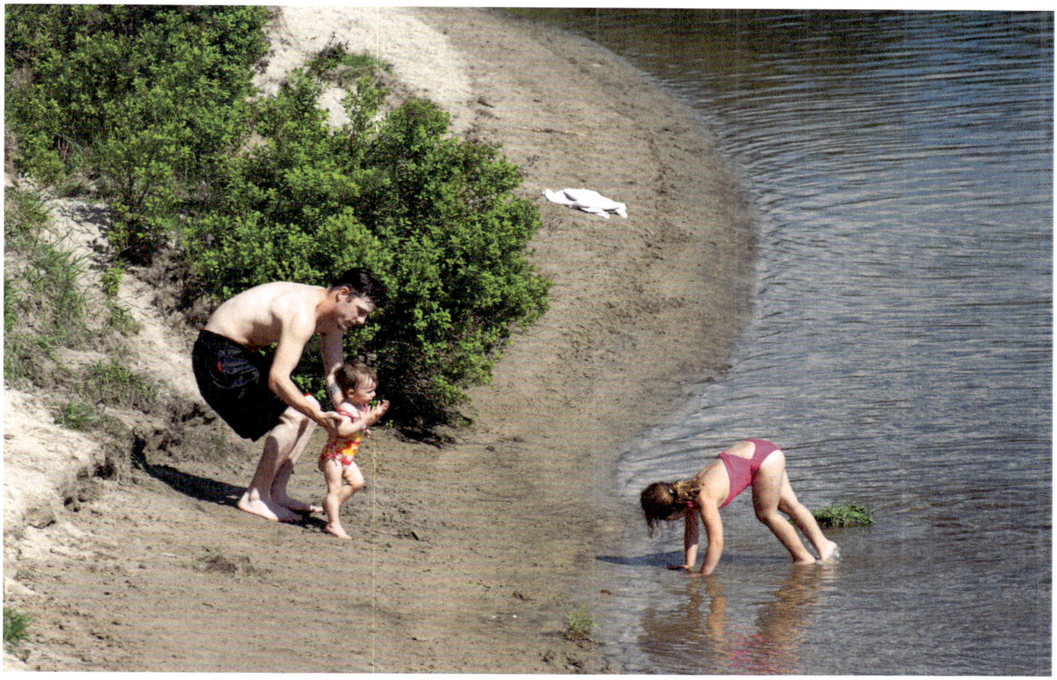

Figure 9-17. *Uncropped*

Here's another example of changing your story. The original version of this family (Figure 9-16) playing on the banks of New Hampshire's Merrimack River one blistering July day really had three stories going on, and they were competing for each other. Look how the various crops change the story completely!

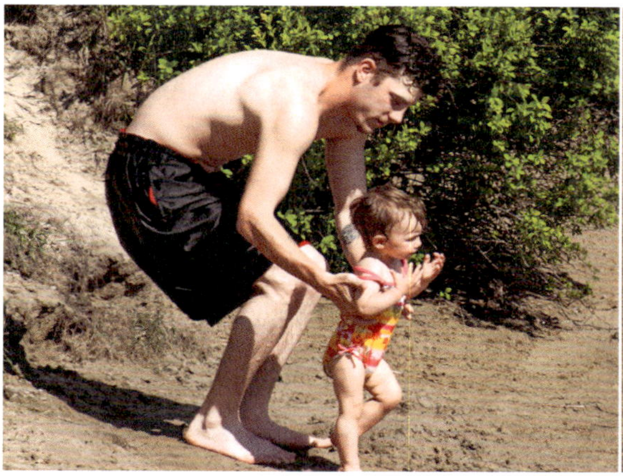

Figure 9-18. *The guy's story*

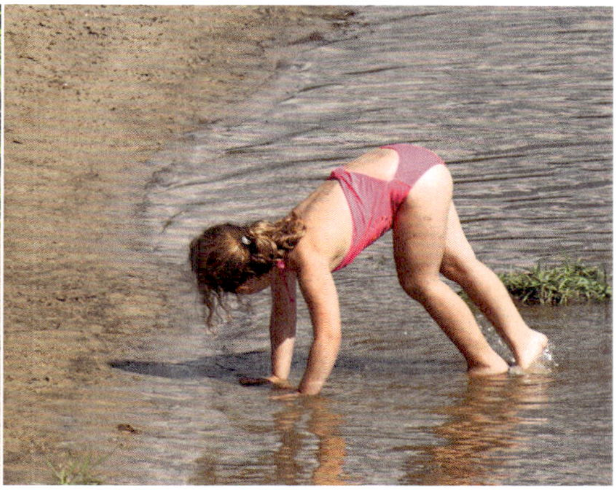

Figure 9-19. *The girl's story*

This image (Figure 9-20) is not quite the same as the first version, Fiigure 9-17, because the crop helps us connect the two subjects by an implied diagonal (top left to bottom right) that you don't see in the original -- too much distraction from the river bank, shrubbery, and the white towel.

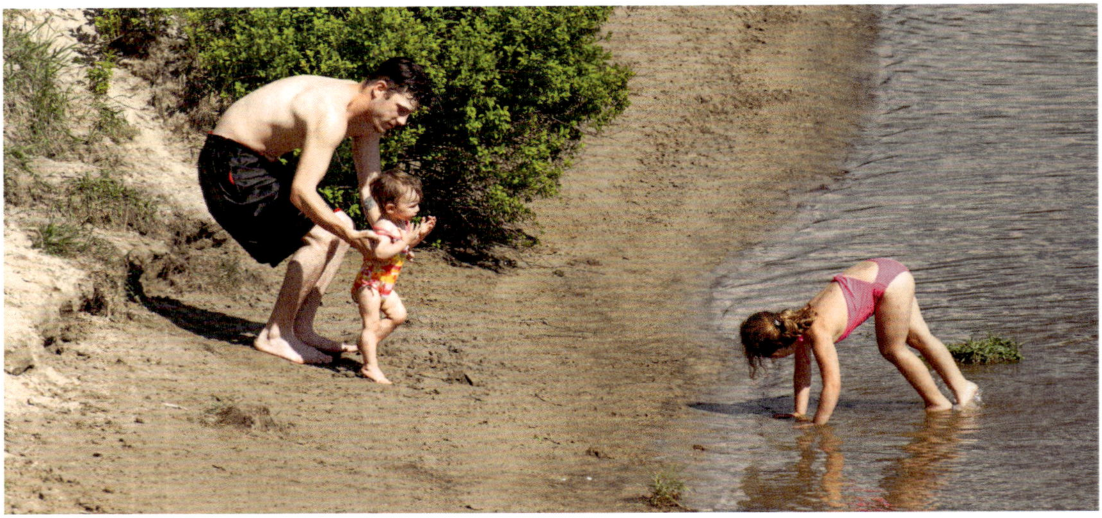

Figure 9-20. *The family's story*

Cropping can improve composition by finding the portion of the shot that most closely corresponds to one of the basic organizing strategies (design, or compositional strategies) for photos. In the case of the cave shot shown above, cropping improved the composition by emphasizing the diagonals in the picture. The same effect occurs in the church window shot. Composing to emphasize diagonals automatically increases the sense of motion and energy in any photograph. See "Composition and Design" in Chapter 5 for more on this subject.

Workflow and Cropping

As I said earlier, the best way to get a picture right is to take it right. But sometimes you simply can't get rid of a shrub, or get to the perfect position to shoot from. Other times you may want to intentionally leave plenty of room to crop the image down to a different proportion, say, to a 2: 1 ratio (panorama) from a 3:2 (standard 35mm format). (On cheap cameras with a "panoramic" setting, all they're doing is discarding the pixels above and below certain pre-set "crop marks.")

Assuming you're going to crop your picture, when should you do it? There are two schools of thought here. One says crop immediately, and then you only have to edit and correct the part of the image you actually wanted. The downside of this practice, especially if you shoot in JPEG mode, is that you will find it difficult to go back uphill and recover lost material if you want it later. If you crop, then save the image, the original pixels are gone forever. Of course if you're shooting RAW, you won't lose the original pixels, but if you crop then edit, then decide to re-crop the original, you'll have to re-edit that copy as well. Tedious.

The other approach, especially if you're shooting JPEGs, says do all your editing, then make your crops. Save your cropped image AS A COPY, then return and undo the crop you just made on your master file, re-crop, save as copy and undo, etc. You don't have to re-edit, but you do run the risk of forgetting to undo your most recent crop before re-cropping, in which case you will lose precious pixels forever.

The best practice is to shoot in RAW (your original data is never touched, so you can't lose it). Do as much of your editing as possible in your general photo editing program, such as Lightroom, Aperture, or CaptureOne. Make your crops from within this program. Then, if you really want to get picky, open the file you want to work on in Photoshop and have at it.

When Not to Crop?

Not every file can or should be cropped. There are both technical and artistic reasons for leaving well enough alone. The artistic reason is if you've done the shot perfectly, don't mess with it – and don't compromise your great picture by forcing it into some other format. Use borders or vignettes if you must, but DON'T MESS UP A PERFECT IMAGE for the sake of a size rule.

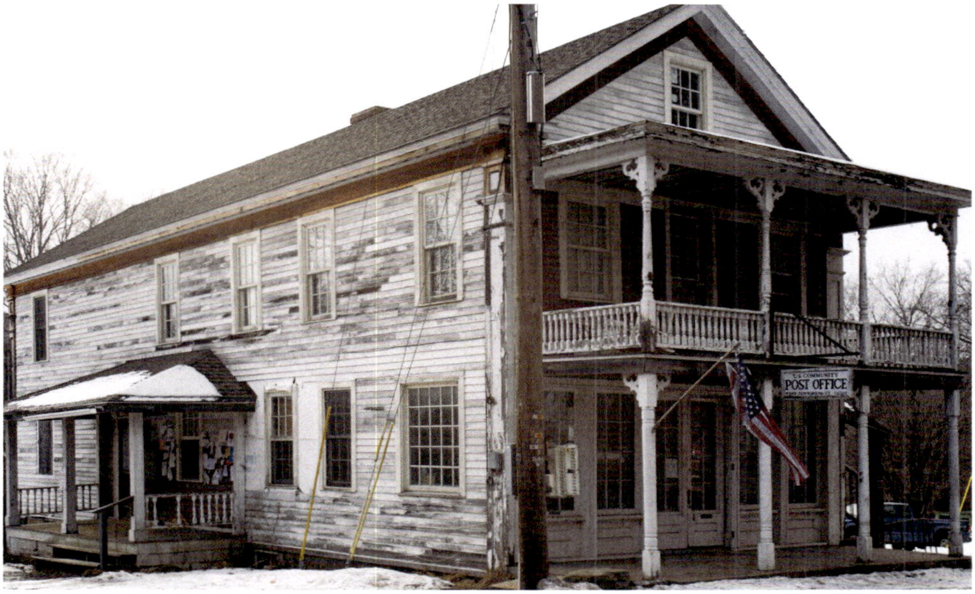

Figure 9-21. *Community post office, West Townshend, VT. Uncropped and not really croppable. The active portion of the shot fills the frame; there is truly no place you can crop without damage.*

The technical issues mainly have to do with prints. Sorry, but no screen image yet made can carry the detail of a fine-quality print. And printing is where you have to pay the piper for your crops.

The technical reasons boil down to "no free lunch." The penalty for cropping small files, or for making VERY tight crops in larger ones, is lousy enlargements. By way of example, a couple wanted me to make an 8x10 from a cheap cell phone photo of their deceased baby. I told them that the picture would be extremely fuzzy, but they insisted I should try. Of course they were disappointed by the result – and I tried every trick I knew for about two hours to get it that far. But you can only do so much. There simply weren't enough pixels in the original image to blow up that big – and cell phone cameras 5 years ago were maybe 2 MP with lousy lenses and produced very compressed jpeg files at that. Even with a good camera, if you crop too far, you will not be able to make good large prints.

At large print sizes, pixels do matter, as does lens quality. The logic of this is simple. An 8x10 has 80 square inches of image. A 4x5 has 20 square inches. If you try to crop a 4x5 chunk from an 8x10 file, then blow it back up to 8x10, you'll have to expand your pixels by a factor of four! That's when things get ugly. So, the more pixels you have in your 4x5 crop, the bigger they will print without problems. To sum up, cropping is one of the best and simplest ways to improve your images. Just don't over do it, and respect the technical issues you may have to deal with. Be careful not to lose your original data; sooner or later it may come in handy!

CONCEPTS IN PHOTO EDITING

Photo editing is as important as preparation, design, and shooting. These days it's known as "workflow," and one hears the phrase "I'll deal with that in post," meaning "post-production," a term swiped, uh, borrowed from the movies.

Photo editing has two phases. First is editing the raw material down to the specified number of pictures you intend to keep. That phase is covered in "Too Much of a Good Thing" also in this chapter. The second phase is working on those images until they are as suited to your purpose as they can possibly be. That's the focus of this article.

Life Begins with RAW

One of the truly wonderful things about digital photography is the number of opportunities you have to experiment, change your mind, or just give up on an image without losing your original image. That's true if you shoot in RAW and less true if you shoot JPEG. If you do shoot JPEG, you can edit your images, but for safety you have to make copies at every stage. Editing a JPEG, saving it, opening it and working on it some more – all this causes irreparable data loss and an increase on artifacts – visual distortions that most often look like wrinkles or smears or frost on a windowpane— keeping your images from ever again looking really right.

If you're not into getting the best photos you can, want your camera to make irrevocable decisions on your behalf, never do any photo editing, aren't interested in making art – well, then, you probably aren't actually reading this, and if you are, I give you permission to skip the rest of the article. And go ahead and shoot JPEGs to your heart's content.

If you do care, shoot in RAW. A RAW image is a file that contains all the data the camera gathered in a particular exposure. You can't even see it if it hasn't been decoded by your software. When you're reviewing RAW images on your camera back, you're actually looking at temporary JPEGs that have a tiny fraction of the original data.

The beauty of RAW is that you never alter the original file until you delete it. It stays pristine, like a princess in her castle, and temporary copies of that file are made in the computer so you can work on them and then save them into whatever format you like.

This means that you can make numbers of different versions of the same file, for any artistic reasons you choose. The original data is preserved like a fly in amber for ever. This is very good news! It also takes disc space, but hey, memory is cheap.

Assuming you've done the culling-out process and identified your keepers, now you can begin making them better. Virtually no images are completely perfect straight from the camera.

Step One — Adjust Color

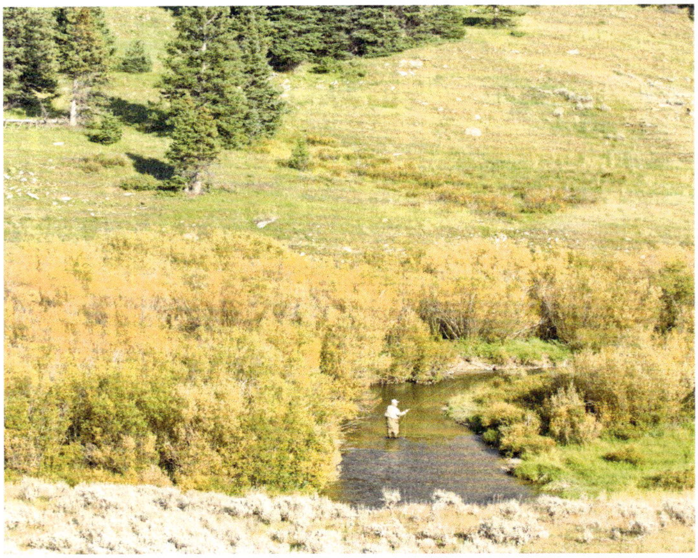

Figure 9-22. Trout fishing in Wyoming. As shot.

No matter which editing program you use, adjusting white balance (color) FIRST makes a lot of sense. Adjusting color affects the brightness of the image, but adjusting brightness doesn't affect the color. So if you do exposure first, then color, you'll have to re-do exposure. Doing color first eliminates this step.

Here is a sample image depicting a fisherman near Burgess Junction, WY. The first picture is "as shot." It's very over-exposed, and the color has way too much yellow and magenta.

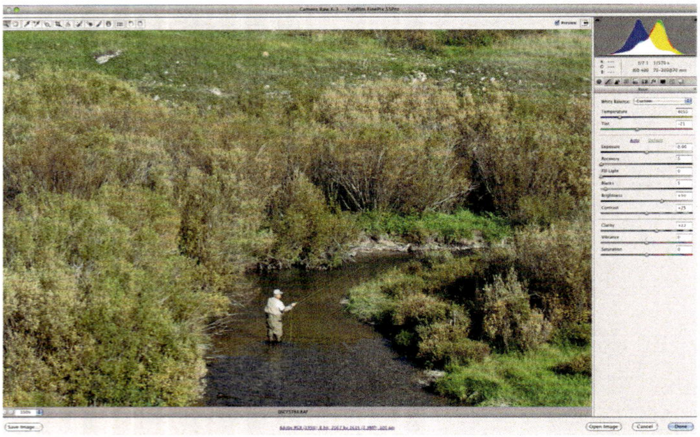

Figure 9-23. Adjust color first

After adjusting color, and only color, you can see that the exposure appears to have improved, as, of course, the color has, too:

Exposure and Curves

Next, adjust the exposure. I typically hit "auto," then tweak from there. In Adobe Bridge (part of Photoshop), the sliders on the right are in the preferred order of making adjustments. Start with color, then work down. You don't have to do it this way, but it's a convenient work-flow. Notice how the sliders on the right in the image below have moved compared to the unadjusted photo above.

One comment: the black slider and the brightness slider are very powerful tools. The black slider moves all the darker shades further to the black, making them denser and more obvious. In an overexposed picture, including many, many landscapes, this is a good idea. But you want to watch not to over-do it and make the picture blocky with an absence of shadow detail. Ditto on the brightness tool. It increases the luminance (light) value of the middle tones in the picture. Used together, they add "pop" to the picture, but again, you don't want to over-do it.

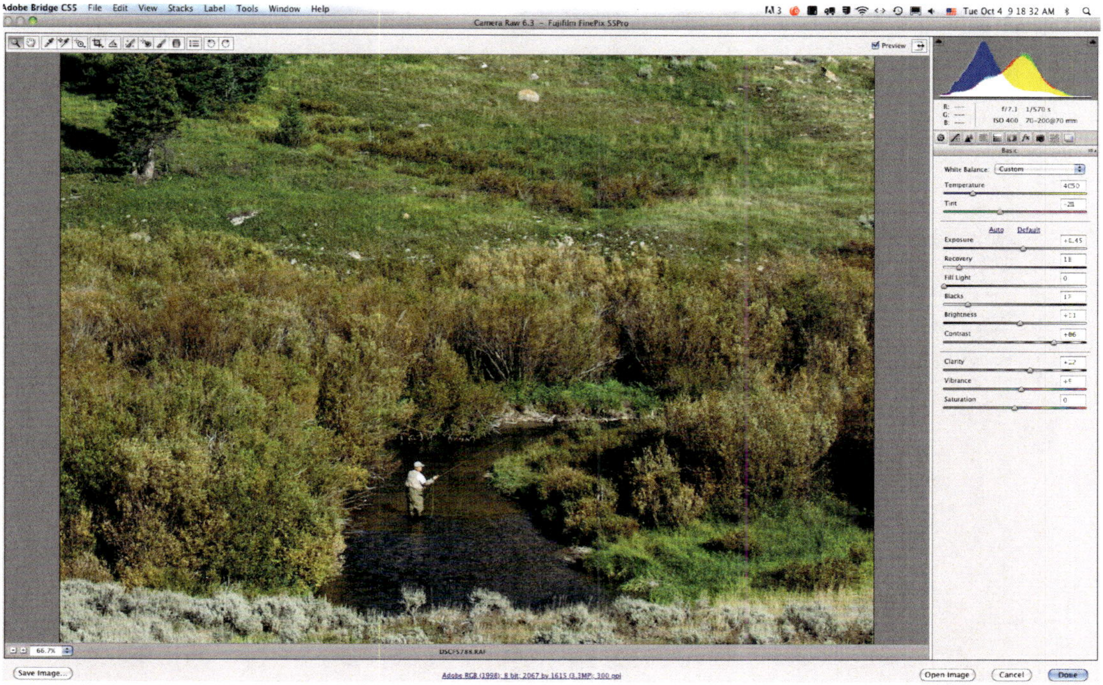

Figure 9-24. *Adjusting exposure and curves.*

From here, I decided I wanted a little more "pop" to the image, but I wanted more control than the black levels and the brightness slider would provide, so I made a slight adjustment to the tonal curve using the "curves" palette. Again, notice the sliders on the right. All four were in the center when

I opened this tool; the adjustment has put a small change in the sliders, but it makes a noticeable change in the image. You can also see that instead of being straight from corner to corner, the line in the grey graph now has a slight curvature.

At this point I would adjust sharpness, remove noise, adjust vignetting (dark shadows at the edges of the frame), remove "fringing" (purple or green lines outside edges in your picture); these show up in "busy" areas like leaves on trees, and most editing software lets you clean up these "fringing" problems.

None of these changes need to be made to the sample picture; it's plenty sharp, not at all noisy (static). Note: when you adjust sharpness and noise, do it at 100% magnification; you need to see what happens to the actual pixels to judge the effect.

You should always view your image at 100% anyway, at least once – because you may very well have dust on your sensor that won't show up clearly on screen at small magnification, but may be really obnoxious when you print the image. So this is the time to use the "spot" or "dust" tool in your editor and clean out the goobers, and catch any "hot pixels" (bright tiny spots of color) while you're at it.

I have lousy eyesight, so I often miss these things. Luckily, I have an eagle-eyed friend, John Mueller, who is more than happy to let me know about them. A second pair of eyes never hurts!

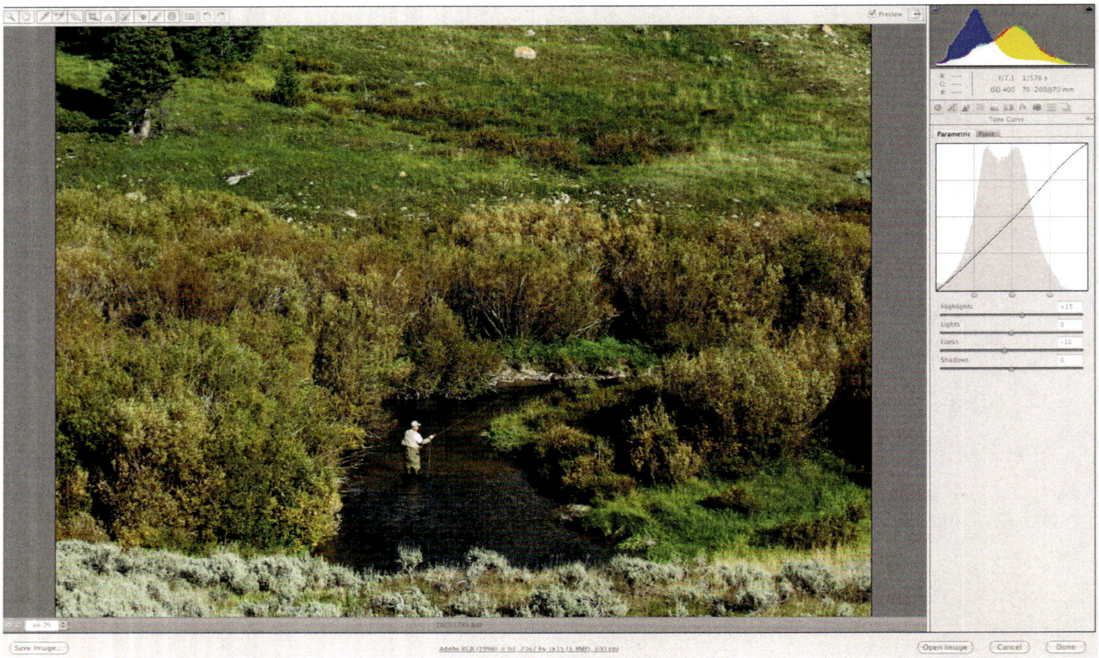

***Figure 9-25.** Tweaking black level and curves using the curves tool in Adobe Bridge*

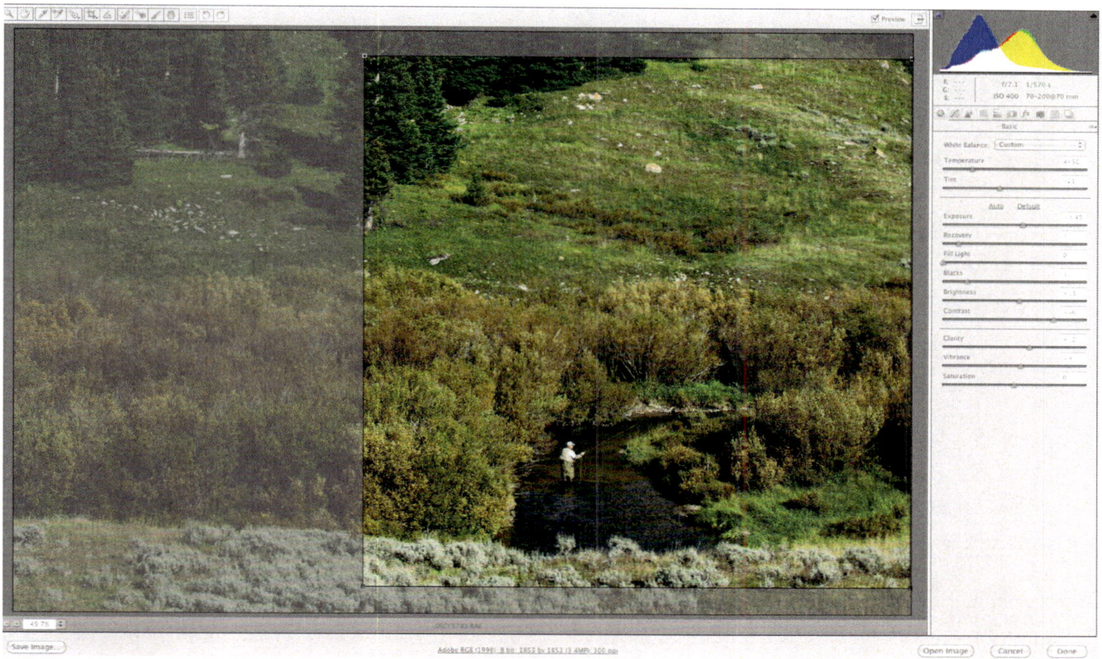

Figure 9-26. *Cropping was the last step in retouching this particular image.*

From here, I have decided to crop the image to change the relationship between the fisherman and the surrounding landscape. I have decided a square crop works best.

So, to summarize, the basic editing sequence is color first, then exposure, then tweak, then crop. You can crop first, if you prefer, but I usually do it last.

I have not intended this to be a tutorial in Photoshop, Bridge, Lightroom, Aperture, or any other image editor. The purpose here is to show an overview of the whole process. For many of you, this will be like mother's milk, but for others it may be new. Obviously, it's a rich subject and you can buy all sorts of books that provide hundreds of pages of detail.

The last article in this chapter, "Diary of a Fine Art Landscape," takes you all the way from thinking about the shot in the first place through a detailed photo editing process.

HDR (HIGHLY DUBIOUS RETOUCHING)

The Old Curmudgeon asks, what is all this HDR crap?

He has a number of answers.

1. HDR (High Dynamic Range) is a series of techniques to assist digital photographers to match the dynamic range of the human eye (depends on who you ask, but that's between 9 and 11 f-stops -- a lot, in simple terms). Digital cameras typically have a range of 6 to 7 stops, max.

2. HDR is used to rescue photos that suffer from camera (or most likely operator) limitations. If the scene is flat as a pancake, lacking highlights or shadows, HDR can help you create them where they were not.

3. HDR is yet another means for the talentless to ape the truly creative people who understand light, contrast, and color.

4. HDR provides endless proofs of HL Mencken's jibe that "no one ever lost money underestimating the taste of the American public."

The Old Curmudgeon may be tactless, but he's not stupid. Still, his less than temperate remarks merit a little elaboration, lest the unwary simply dismiss him with a shrug.

HDR is in fact a procedure to create a dynamically rich image where the original RAW interpreter saw a very limited image. In short, it's a method of making a silk purse out of nature's occasional sow's ear.

Today there are a ton of gimmick-laden HDR applications. You don't need them to achieve a good result – especially if you shoot in RAW. If you shoot in JPG mode, you're cooked, so fuggedaboudit. (You may THINK you're OK, but then look at your image at 100% size, that is, 1 image pixel = 1 screen pixel, and you'll soon see how bad things look). All you really need is one more or less correct RAW exposure. Make three copies of this exposure, setting the exposure control to -1, 0, and +1 stops (or more, depending on what you want to accomplish). Open all three images in Photoshop, copy them as layers to a master image, then erase as needed to get the enriched image you wanted. (You can make lots of other adjustments one layer at a time, but this gets you most of the way there.)

This is oversimplified, but not by much. What this basic technique allows you to do is to put details into shadows and bright areas where there were none before, as well as make the darks darker and the brights brighter.

You get better results if you take multiple bracketed exposures of the same scene, knowing you'll be using HDR techniques back home. Use three or more registered exposures 1 stop (or more) apart.

Here are three examples. The first is as-shot, the second is the very best HDR Photoshop could produce, and the third is hand-adjusted. Of course, shooting three registered exposures was impossible under the circumstances of bouncing around in a Cessna at about 120 miles an hour, so I made do with the one RAW image I was able to capture.

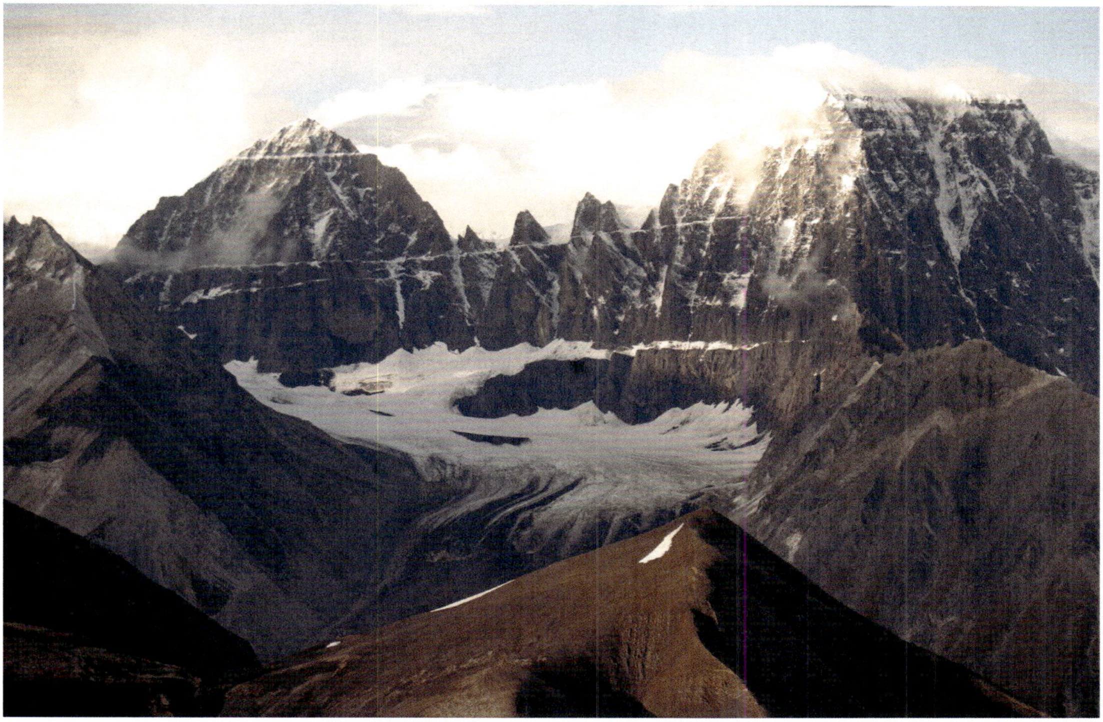

Figure 9-27. *Aerial view of cirque, central AK. As shot.*

The original is overexposed, but simple Photoshop / Bridge adjustments don't take care of the haze and contrast problems caused by misty air and a plexiglass window. The HDR version is over-dramatized; it may look OK here, but it's disastrous at 11x14 or larger. It's the photographic equivalent of a wet towel flung at your face.

The hand-altered version, prepared using the method described above, could be tweaked a bit further, but it clearly solves the exposure problems, leaving a highly dramatic photo that maintains a more natural appearance.

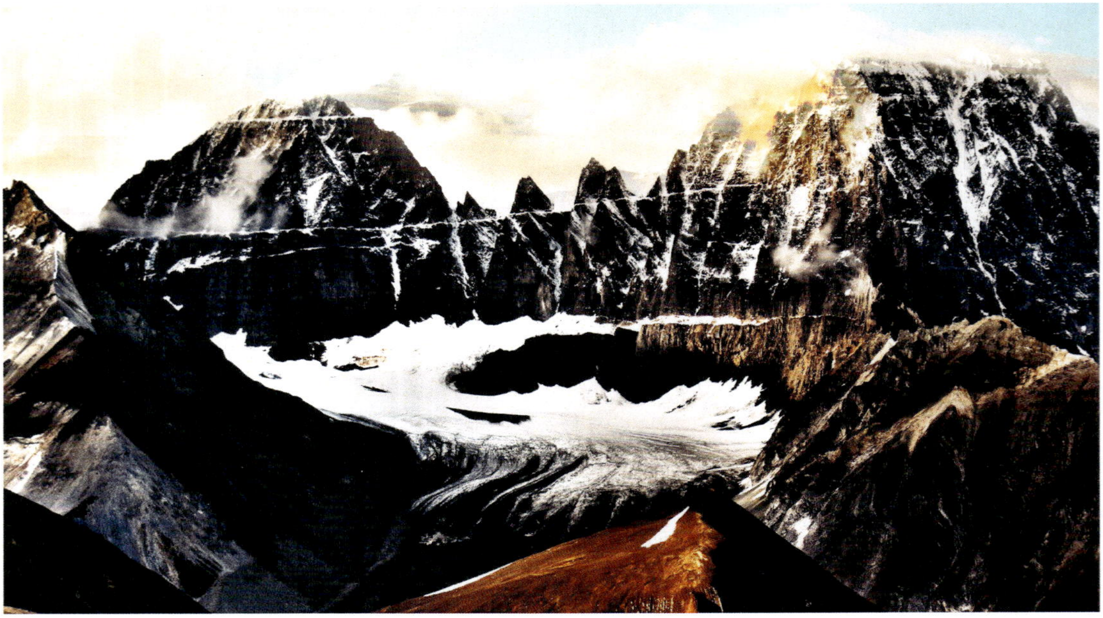

Figure 9-28. *Best result using Photoshop CS5's HDR routine.*

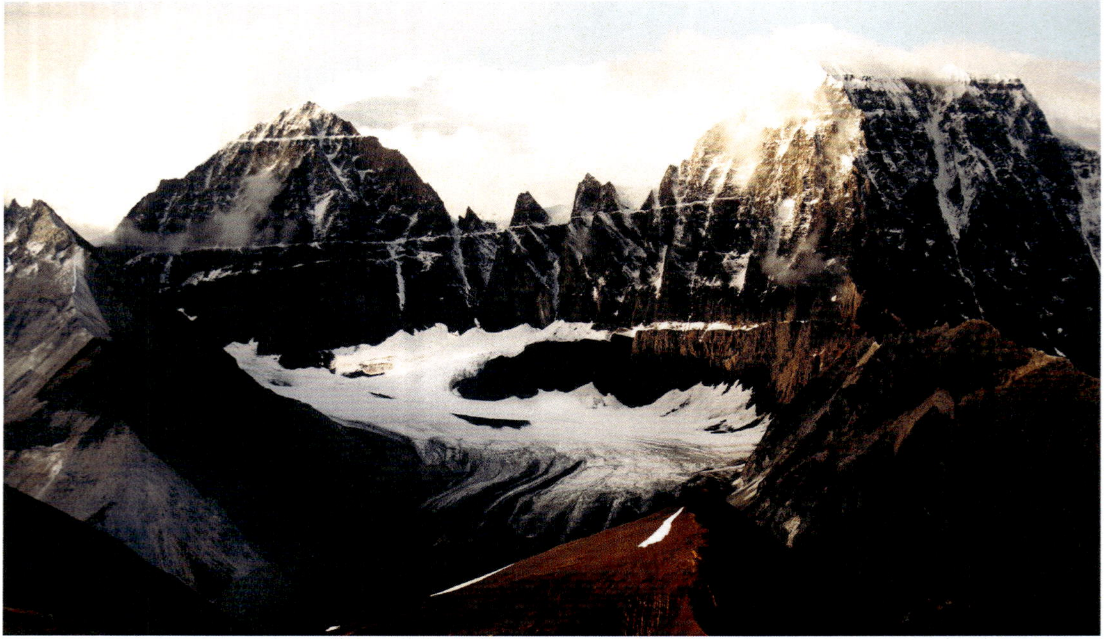

Figure 9-29. *Hand-done in Photoshop CS5. Chief differences are in color saturation and not pumping up the yellows.*

This is a vital point. To use HDR or any image-manipulation tool effectively, you HAVE TO KNOW WHAT YOU WANT TO ACHIEVE. I don't mean you have to put it in words ¬– but you have to have a vision of what will be the right treatment for THIS image, what will best express the vision and feelings you had in your mind when you took it in the first place.

I'm not trying to discourage folks from trying HDR. By all means, tinker to your heart's content, learn the plusses and minuses, the ins-and-outs of this technology. But please, please, DO NOT thrust your baby steps on an unwilling audience, and then expect praise. At the very least, have an idea in your noggin of what you want to achieve, and how you want people to respond to it BEFORE you play with the buttons, and ABSOLUTELY before you post an over-colored dish of M&Ms and try to persuade us it's an orange soufflé.

HDR apps invite the most wretched excesses and encourage people to rely on technology when their inner eye has no clue as to what they want to achieve. But -- and this is important -- the problem is not in the apps. Yes, they invite pictures that make the Old Curmudgeon run for a barf-bag. But they don't have to be used that way. Like guns, or helicopters, or motorcycles, they can be used for good or ill. Like any tool, they depend on the judgment and skill of the user to be effective. To be blunt, James Cameron achieved HDR / CGI miracles in Avatar, but very few of us are James Camerons!

Properly used, HDR can salvage deficient photos, enrich good ones, or bring a spot of magic to an otherwise humdrum shot. But improperly used, HDR is yet another reason why modern photography is so often held in contempt – and so little purchased.

The Old Curmudgeon realizes he's an old dog barking into the wind. The Nikon D800 (just released at this writing) will take two immediate exposures as far apart as you specify (up to 3 stops apart, which is a lot) and combine them for you into an HDR image, right there in the camera! The Old Curmudgeon recognizes when the jig is up.

The Old Curmudgeon concludes his rant saying, "Take good photographs first, based on your eyes and your heart and your skills. Feel free to see what technology can do to help you get there. But if you rely on HDR apps and autopilot to do it FOR you, you will never get anywhere worthwhile." Amen.

CASE HISTORY OF A FINE ART LANDSCAPE

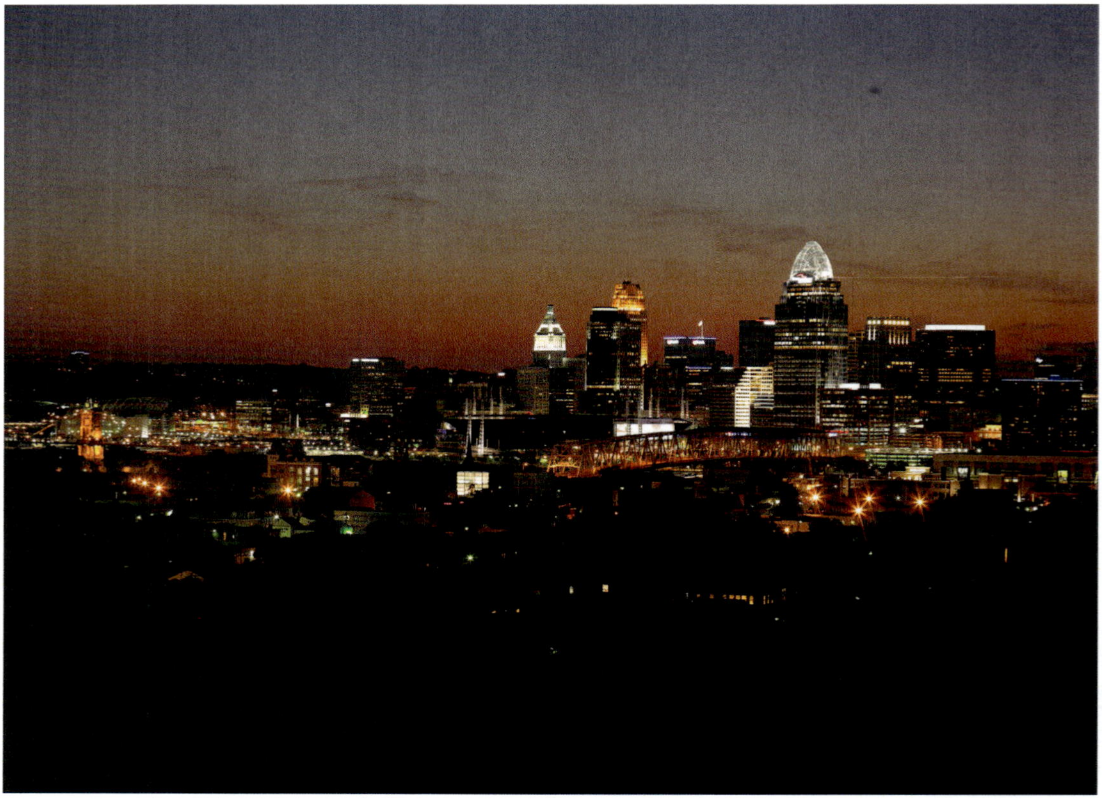

Figure 9-30. *Cincinnati Twilight as shot, seen in Adobe Camera Raw (Bridge)*

My friend and fan, Reggie, saw a copy of the Cincinnati twilight image above and asked me to tell him everything that occurred from taking the picture through final processing.

At first I thought that might be easy, but as I thought about it more, I realized that there were far more steps involved than I had thought. While I don't produce any polished images on autopilot, there are so many routine steps (which can vary in sequence) that I've never listed them out. Perhaps it's time I did. Hence, this Diary of a Fine Art Landscape.

Concept

The photo really began about the middle of May, when a commercial client of mine said that he would really like to have current pictures of the Cincinnati skyline. In the past year or so the skyline has really changed; now it's dominated by a tall building with the "cage" or "tiara" on top. It's supposed to be the crown on the Queen City of the West (as Cincinnati was known up til 1875 or so).

Cincinnati looks pretty pedestrian during the day, but at night it takes on certain magic. Unfortunately, in the middle of summer the weather does not help with landscape photography. The problem is haze. I wanted a picture that would be clear, sharp, and with good sky color behind the buildings. This meant waiting for a clear cool day in the summer time, which in Cincinnati means waiting for a cold front to come through, or for hell to freeze over, which ever happens first! It also means timing a shot so there isn't too much moonlight; it tends to wash out the sky.

Finally, on July 14, the magic weather conditions occurred. In the late afternoon sunlight my friend, Tina Karle, joined me to shoot on the Kentucky riverfront, and then we moved uphill to the Newport Catholic Center. Since I was last there two years ago the view of Cincinnati has been narrowed by new condominiums and by trees which have grown to block some of the field of view. But there was still a very narrow space of lawn available that gave a good view of downtown Cincinnati.

Setting Up

We positioned ourselves carefully to avoid reflected light pollution from the lawn-mounted searchlights that illuminate the building after dark. The next step was getting the tripod stable and dead level. If you're thinking of stitching images to make a panorama, getting a level camera is very important. It avoids distortion and gives you the maximum usable pixels to work with.

Once the camera was set up it was really a matter of experimenting with lenses exposures and making sure that the camera would have as little vibration as possible.

Vibration Reduction

Here's the deal with detail-loaded landscape shots that you expect to print at very large sizes (like 30 x 60 inches). Any form of vibration kills you. You cannot be too careful about this, because it's the limiting factor between a great shot and a mediocre one. What happens is that even the smallest camera shake – invisible at normal sizes – creates a blur at very large magnifications. Imagine you have a 6 x 4.5 cm sensor (medium format size). That gives you a "negative" that's 27 square centimeters, or 4.18 square inches. I don't care how many megapixels you've got, if you blow your 4.18 in "negative" up to 24 x 30", you now have an image that is 720 square inches. That's a little less than 175 times bigger than the original "negative". If you blow ANYTHING up by a factor that large, the least little thing that's wrong is going to look monstrous in the final print.

Camera shake isn't the only variable. The size of the sensor, camera shake, vibration from wind, lens sharpness, focus, atmospheric haze, heat "shimmer," and ground vibration all matter if you are shooting with wall-size prints in mind.

Exposure time increases the risk of camera shake. This shot was a 10 second exposure, during which NOTHING could vibrate.

I took the following steps to minimize vibration:

- Good tripod, rated for my 5-pound camera.
- Solid footing for tripod.
- Pre-focus (manual)
- Camera mirror pre-up (so it doesn't cause vibration when you take the shot)
- Self-timer set to 3 seconds to give thing time to settle down.
- Use a remote shutter release if you have one; this camera doesn't
- When you press the shutter, take one or two steps backward (gently) so your lead feet don't cause camera shake during the exposure
- Don't move til you hear the shutter click shut

Camera and Lens

This time out I was using a Mamiya 645 AFD medium format camera with a Phase One P25 (22 megapixel) back (long since obsolete, but still does a great job). I used 3 lenses: a 35 mm wide-angle (about 20 in SLR speak), a 55-110 f 3.5 (about 30 to 65 in SLF terms), and a 120 mm manual focus lens, one of the sharpest lenses ever made. This particular shot was made with the 120 MF, exposed at ISO 100 for 10 seconds at f8.0.

Exposures

The selected shot was one of a series of 4 exposures shot at varying exposure times with an HDR experiment in mind. This was the best overall exposure, but despite the 9 stop dynamic range of the P25 back, there were some issues to be dealt with later in the computer. At that exposure length, brightly lighted highlights are going to lose detail, notably the brightly illuminated signs and the floodlit "tiara" on top of the Great American Insurance building. Note: the light was changing very rapidly, when twilight was turning into night. So I had to keep metering, or at least varying, to adjust to changing conditions.

I used – gasp—a light meter (Seconik 350) to get an overall reading. The actual exposure was ISO 100, f8, 10 seconds.

I chose f8 for maximum sharpness. If I'd shot at f5.6 for 6 seconds the image might have been less sharp and would have had more shadow detail, but you have to choose your enemy: highlights or shadows vs. sharpness.

ISO was set to 100. This particular digital back is an older one and has some serious limitations newer models don't have. Theoretically you can shoot at ISO 800, but you are dealing with massive noise. I never use this camera at more than ISO 200. The intention is to limit noise, but the price you pay is longer exposure times. If I had another $20,000 to spend, I'd get the update. Hah!

Color balance

I shot in "daylight" mode. This preserved the blues, but needed a minor correction later.

Back in the Dark Room (Computer, of Course)

I use a combination of post-processing tools. Adobe's Camera Raw does a truly lousy job interpreting Phase One raw images: Camera Raw thinks you've underexposed every shot by at least 2 stops.

Accordingly, I use Capture One Pro, version 6.2. This software is a pain in the butt (for me at least) in downloading and selecting files, but it is an absolute champ at displaying and editing RAW images ... including those from Nikon, Canon, etc. The current version gives you all the local editing techniques of Bridge or Lightroom, plus it gives you the ability to store variants and output them to various formats. Most amateur shooters don't need this, but remember, my purpose is to produce terrific HUGE prints.

Once I've done most of the editing in Capture One, I process the file and save it as a TIFF file, original size, 300 PPI. From there I will open it in Photoshop and make final corrections, including things like removing the streak of crud in the upper left corner of the image–I don't know if it's a jet contrail or crud I couldn't get off the lens.

Editing Workflow in Detail

After setting up my various working folders (select, output) I go to work on the image. In general, one should tackle color balance first because it affects exposure and other variables down the road. Here is a screen shot showing the unretouched file, with the original color balance. The next shot shows the color re-balanced; in this case I just went from daylight to P25 product flash – and presto, all warmed up, as in the next screen shot.

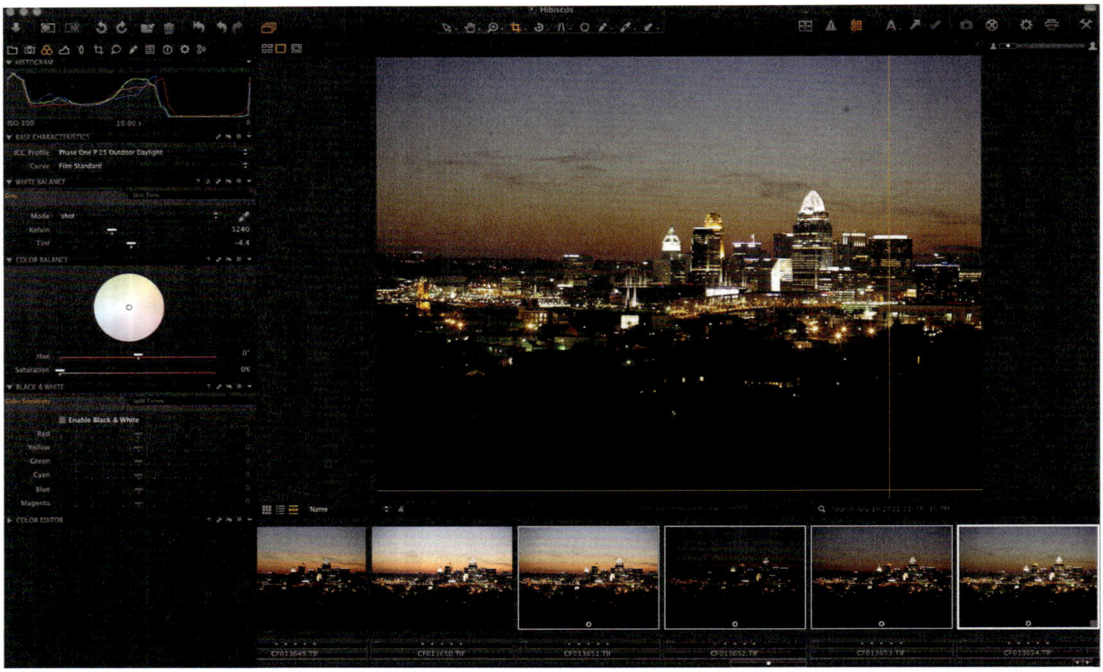

Figure 9-31. Color balance as shot.

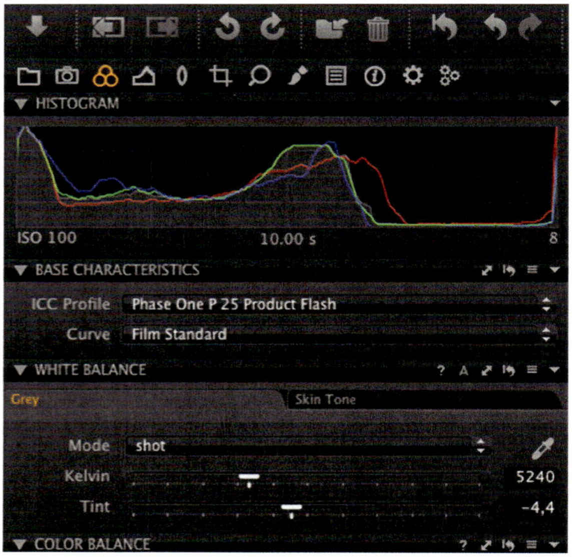

Figure 9-32. *Color correction brings up reds a small amount.*

You can also see the histogram in these screen shots. This is a fundamentally dark image, with a tiny cluster of bright pixels all the way to the right side. Reds are actually not very bright colors (see "What's Red and Black and Out of Focus" elsewhere in this book). Blues are dark, so even though the image looks bright enough to the eye, the camera doesn't see most of this scene as much above middle gray. You can't auto-correct this kind of picture because the software will seek to balance the image – and this one is intentionally NOT balanced. But those tiny bright spots are blown out, making it hard to see detail in the glare of the very bright lights on the Central Trust building and the Tiara.

Exposure Correction

Moving to the next panel, I've made only very minor changes to this picture. I wanted to be careful to get good contrast but not too good, and I moved the black point in the curves window. In the HDR panel I did dial down the very bright areas (the bank tower and the tiara), but more work would be done on those problem spots later.

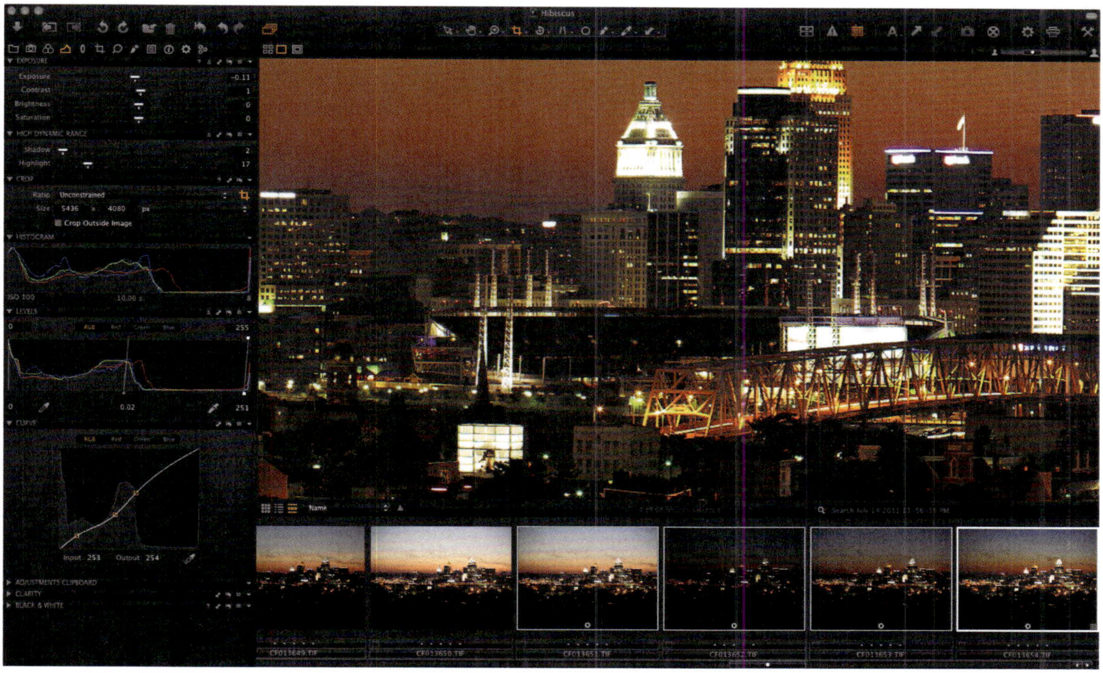

Figure 9-33. *Correcting exposure. Notice that the warmer color tone in this version reflects the color correction made in Figure 9-32.*

The next step was to straighten the verticals. Despite all my care, the camera wasn't 100% level, so the verticals were out of plumb by about 1/2 a degree. Not much, but I wanted to fix it.

Straightening Verticals

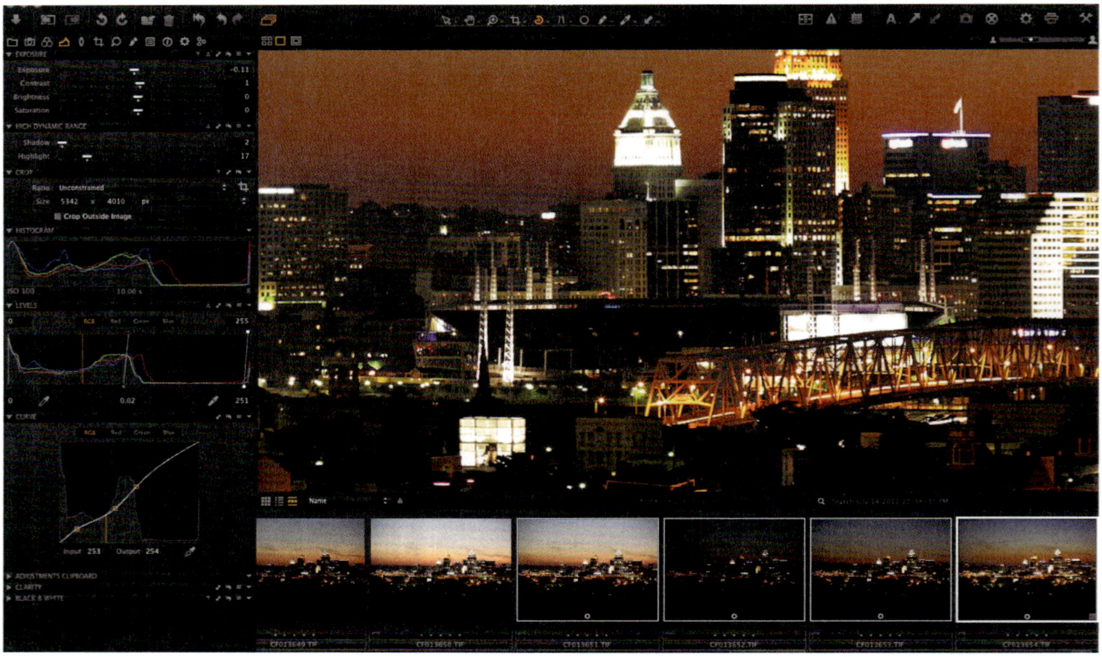

Figure 9-34. *Straightening vertical axis*

Removing Dust and Crud

The next screen shot shows dust elimination. You can do this in Lightroom or Bridge just as well, but it's easier in Capture One. In Photoshop, you can just use the spot healing brush.

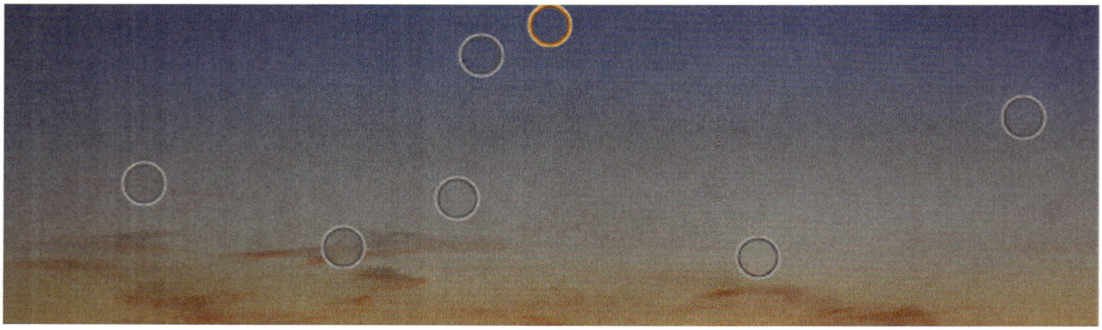

Figure 9-35. *Dust and spot removal*

What I did NOT do in CaptureOne was sharpening. That gets done at the very end of the process, when the image is set to its final size.

I saved this version as a hi-resolution TIFF file, and also saved the other three files in the set

Over to Bridge / Photoshop / Advanced Editing

I opened the file in Adobe Bridge because I wanted to play a little with the color. Bridge lets you control the luminance (brightness) separately for a number of different colors, including red, orange, yellow, green, blue, magenta, and purple. You can also control hue and saturation for those same colors. For this image, I wanted to tweak the luminance of the orange, but in the end I left it where I found it in Capture One.

I decided to get cute and do something about the blown-out highlights. So I opened a program called HDR Express and built an HDR composite. Saved this, and opened it in Photoshop.

Next, I opened the master image in Photoshop. Its as-shot size is 18 x 14 inches (I produced the web image straight from this version). I decided to make a panoramic version 18 x 9 inches. Again, no extra sharpening at this point.

Blown Highlights

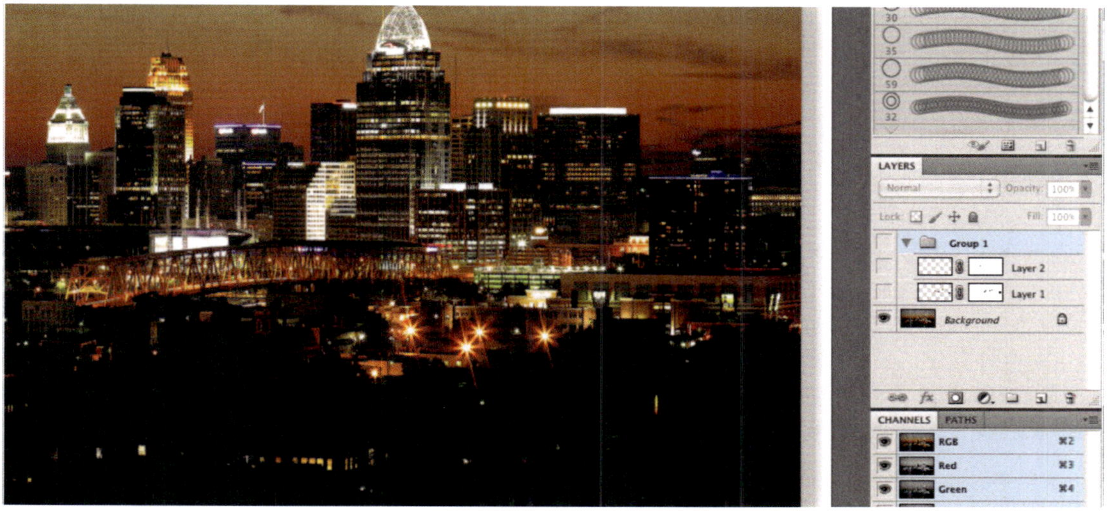

Figure 9-36. *Repairing blown highlights by patching in matching, but unblown, elements of another exposure.*

I made a loose selection around the areas of the HDR shot that had been bothering me, pasted them into the master as a new layer. Cut the opacity to 40%, repositioned each element (I hadn't straightened the verticals in the HDR composite, so had to tweak the imported materials to get 'em perfectly aligned. My bad!) BTW, the HDR version of the shot looked lousy (as it too often does in my opinion) but it was kind of a nifty way to handle this particular problem.

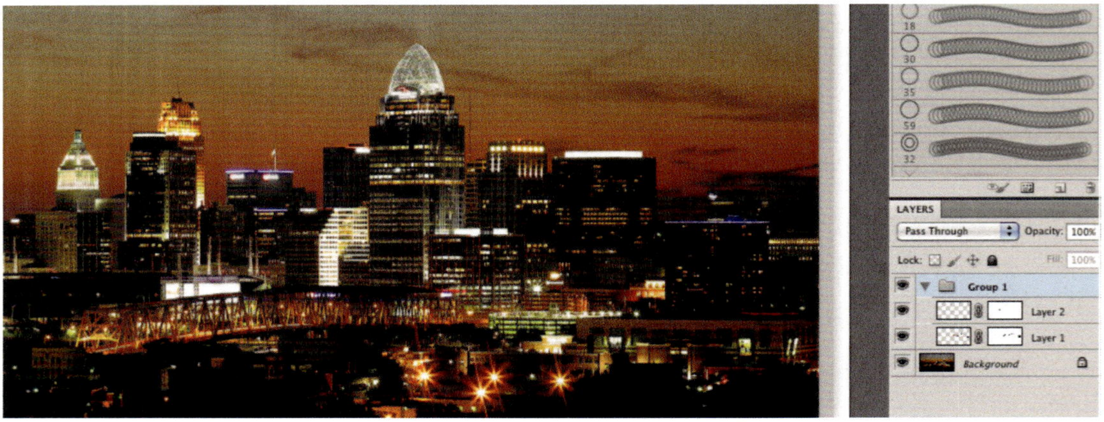

Figure 9-37. *After fixing blown highlights*

Sharpening, Noise Control, and Resizing

I could stop here with this picture. It does not seem to suffer from excessive noise. It's sharp enough for all normal purposes. So why not quit? I said at the outset that my goal is to produce very large prints at very high quality levels. In this league, 9 by 18 is a starter kit! I'm pretty sure I have a client who would love to have this picture at 15 x 30, and to do that requires some extra steps.

I have learned very painfully that if you want to enlarge an image to really big sizes, you absolutely should not sharpen it first! Even using the very good sharpening routines in Capture One and Photoshop, once you go into enlarging the image you affect the sharpening already done, giving a very grainy, over-sharpened texture to the picture. Over-sharpened edges look really hard even though they may not have halos, and the overall effect, while painterly, isn't always appropriate. To avoid this problem I use a program called Perfect Resize. In an earlier life this was known as Perfect Fractals.

The next two images show sharpening off, then with sharpening applied.

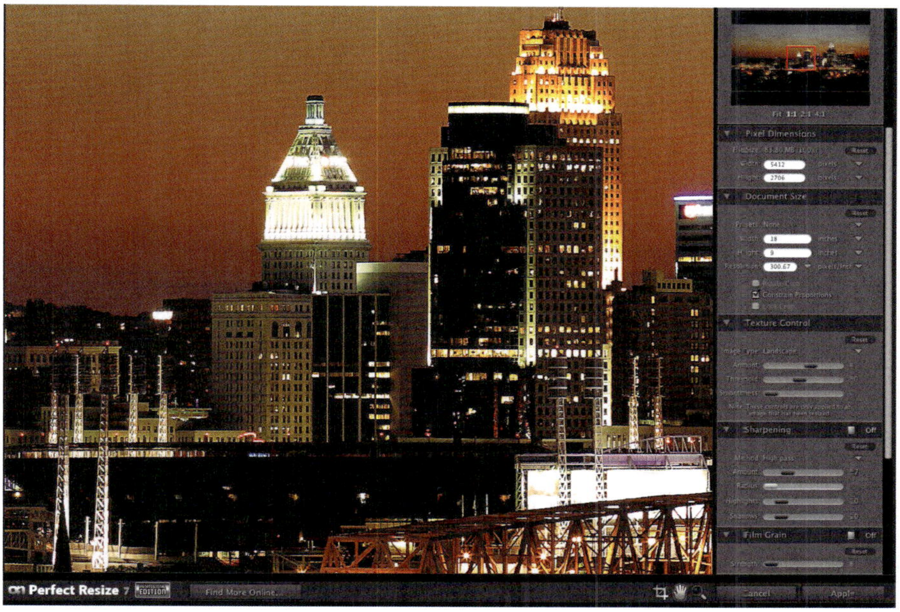

Figure 9-38. *Unsharpened*

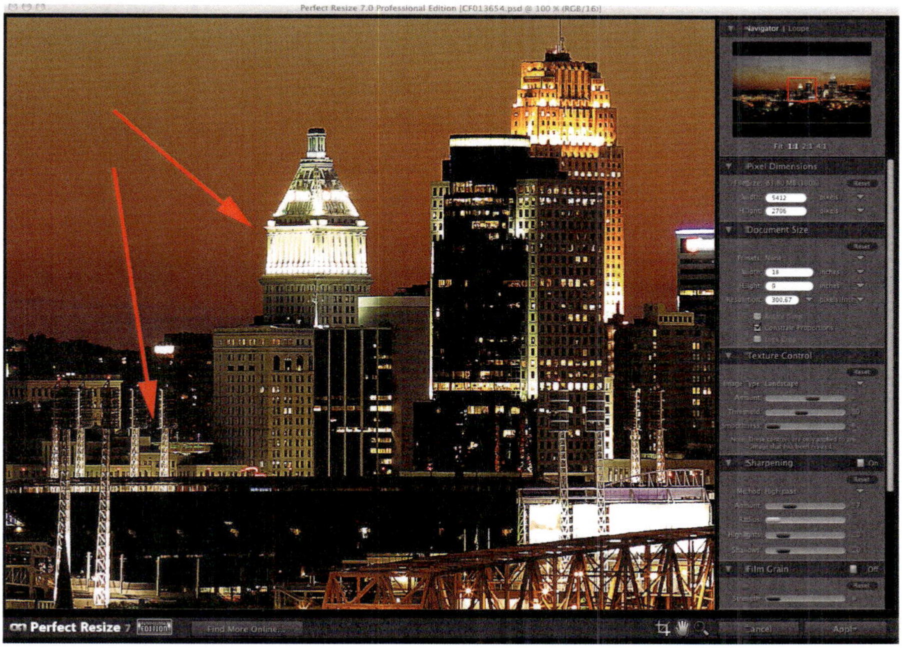

Figure 9-39. *Sharpening applied. Notice the detail in the white cupola and the clean edges on the antenna array.*

Perfect Resize has three different sharpening routines built into it. By experimenting, I have found that even if you don't use it to do its main job of enlarging, it does a very good job of sharpening. The two screenshots shown above illustrate that. These shots show a selection from the picture at actual pixel size, that is, 100% magnification. The image, however, has been left at its 9 x 18 size. I will have absolutely no problem getting a superb print at 15 x 30. I know this because I have already pulled the print 25 x 50 from the same series of pictures, and it looks terrific!

There are lots of different ways to control noise. In the days of film noise was the equivalent to graininess in the film, and the faster the speed of the film the more grainy it would be. Nowadays, the higher the ISO-value you shoot at, the more noise you have. If I had shot this picture at a higher ISO value or for a much longer exposure, it would have had some noise in it.

Once I've enlarged the picture to its final size and included sharpening, I take look at the image in Photoshop and see if extra noise crept in. If it has, I run a noise reduction program to take care of it. The one I use is Noise Ninja. I find that it does a good job overall, and if you need to tweak the controls you can do it. This particular image doesn't seem to me to have significant noise problems even at 15 x 30 so I'm not going to run subsequent routines on it.

There is an argument about when you should reduce noise. Noise reduction works by applying a small amount of blur to grainy areas of the image, then re-sharpening them using unsharp mask to keep them from being too blurry. Some people say you should do this before enlarging and some people say do it after. I tend to do it afterwards. In this case, noise was not an issue so I didn't bother with noise reduction.

So, as Reggie requested, that's the soup to nuts account of the creation of this image. I know it will be too detailed for some and not detailed enough for others, but I hope you have found it useful. Here's the finished picture -- definitely not full sized.

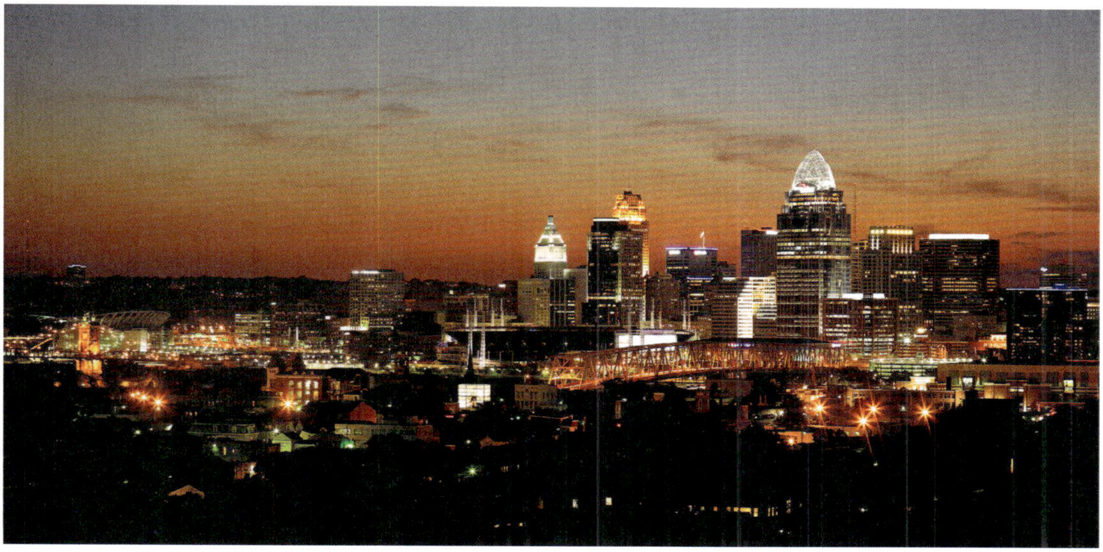

Figure 9-40. *Final version.*

CHAPTER 10: *Photo Restoration*

BACK FROM THE DEAD
WHAT TO DO WITH WATER DAMAGED PHOTOS
A LITTLE GHOST STORY

CHAPTER 10: *Restoration*

BACK FROM THE DEAD

You would not believe the number of customers who come to me because Walmart (or a web-based restorer) couldn't deliver the goods. Then they're shocked when I tell them what the work will cost. So I thought that I'd walk through a photo restoration project, illustrating the process and talking about some of the alternatives one could choose.

Here's the original. The "active area" (the part the customer cared about) was only 2.5 x 1.85 inches in size. As you can see, it had faded to near-invisibility.

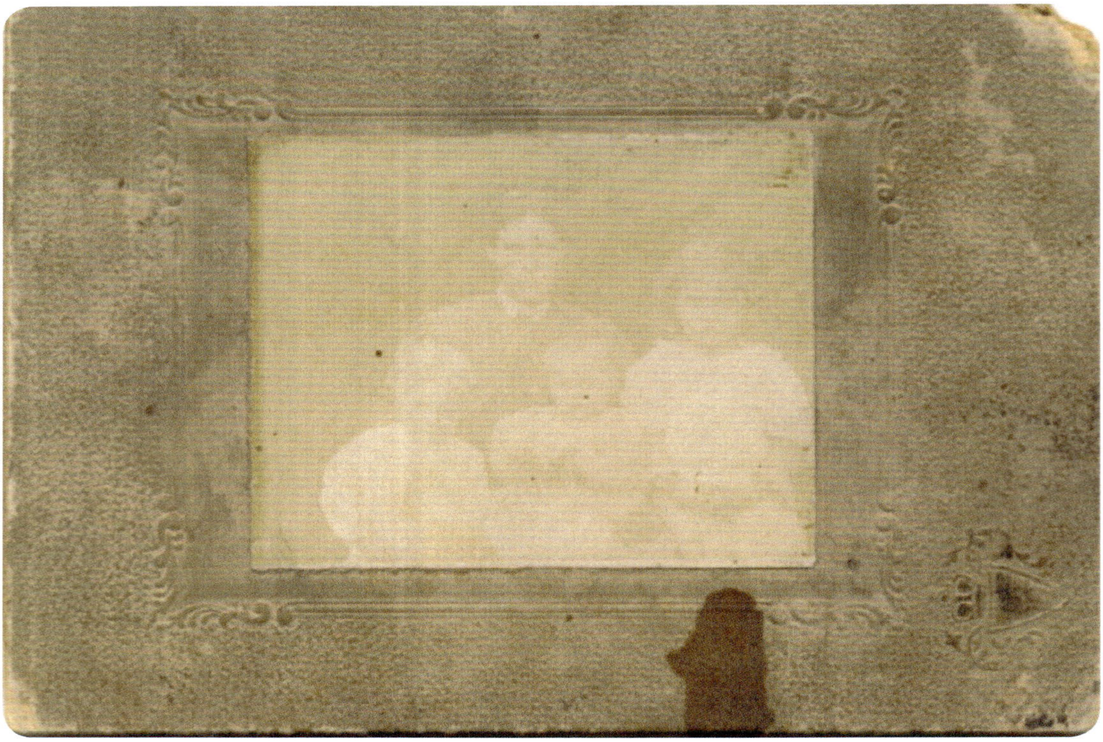

Figure 10-1. *Client's original*

The first step is to ask the customer some questions. "Tell me about this picture" is a great one to open the conversation. It encourages the client to tell his or her story about the people in the picture, how she came into possession of it, where it was hidden for all these years, etc. "What would you like to see as the result of the project" is another good question. "How big would you like it" is a key question, too. Inevitably people want their pictures larger .... but frequently they don't understand the relationship size has to difficulty, or the limitations of the process.

Once we've set goals for the project, and I've looked at the job under an 8x loupe (and shared that with the customer) to make sure I know what I'm getting into, I quote a price and begin the work.

Projects like the one shown here are traps for me --- and I fell into one this time. When the picture has faded as far as this one, and you create enough contrast to see the detail, you create huge numbers of nasty little artifacts. Just how bad these turn out to be governs a lot of the time it takes to clean them up. In this case, I underestimated by 60%. My loss. I scanned this image at 1200 ppi, knowing that I'd be able to get a 5x7@300 out of it, and still have some headroom to work with.

Here's the result of the initial scan (the border came afterwards, but it's irrelevant at this point):

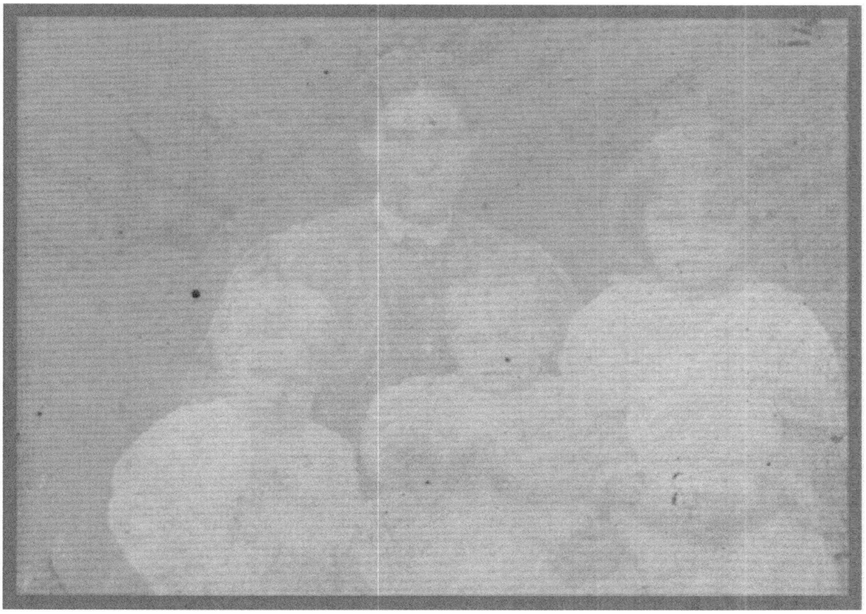

Figure 10-2. *Color stripped out and image cropped*

As you can see, there's still not much to build from. So up goes the contrast via levels adjustment:

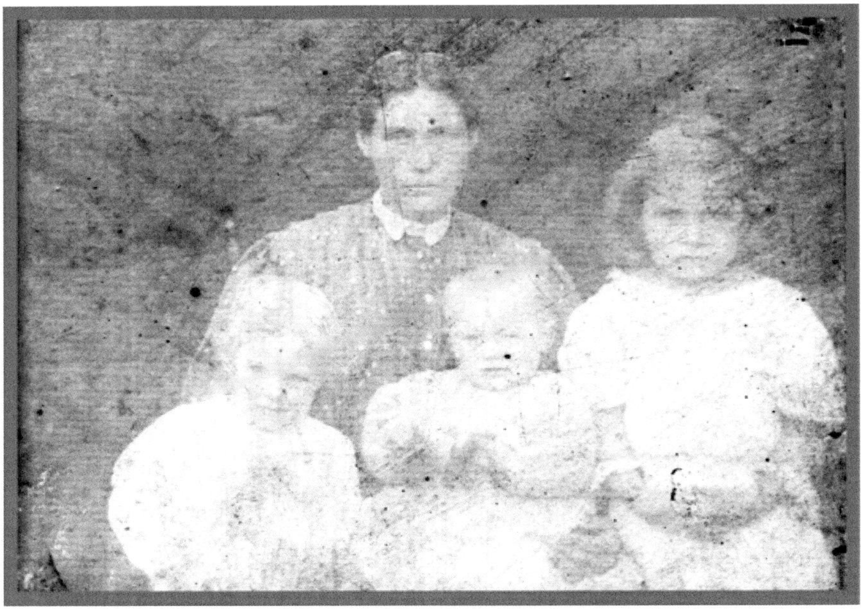

Figure 10-3. *Increasing contrast and luminance reveals blemishes*

The levels are better, and you can see the faces, but the image is still washed out AND grainy.

The next stage was to manipulate using various layer modes, trying to get the best combination of contrast and good texture. But the noise and artifacts kept getting worse. I tried Noise Ninja, but you can't blast the noise without killing too much detail. So it's an interactive process leading from Figure 10-4 to Figure 10-5.

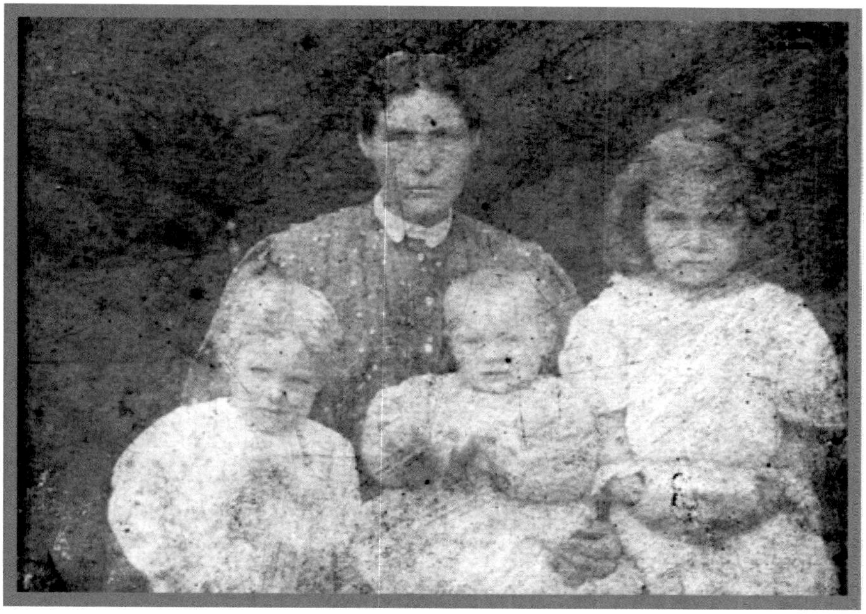

Figure 10-4. *Removing noise is a multi-step process in this case.*

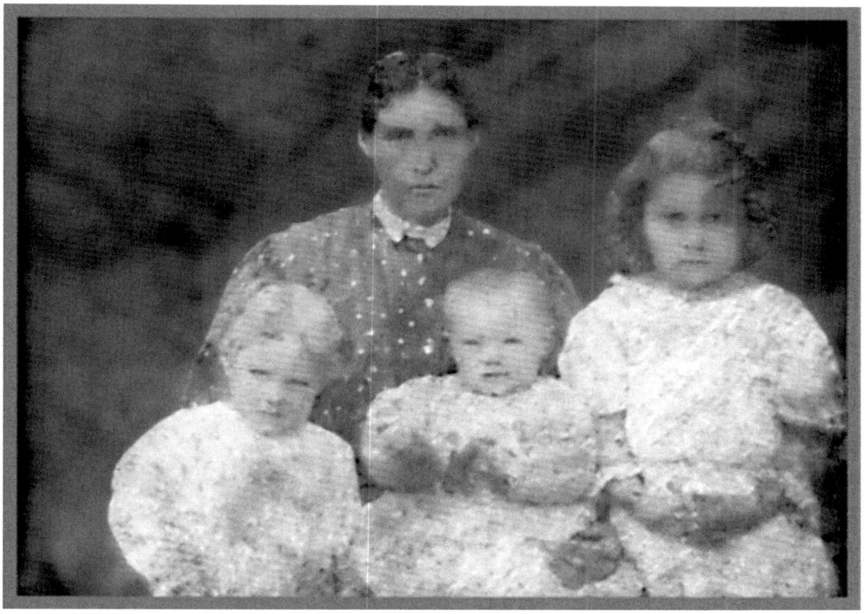

Figure 10-5. *Background cleaned*

To reach the version immediately below , I isolated the background and put the people into a separate layer, Figure 10-6.

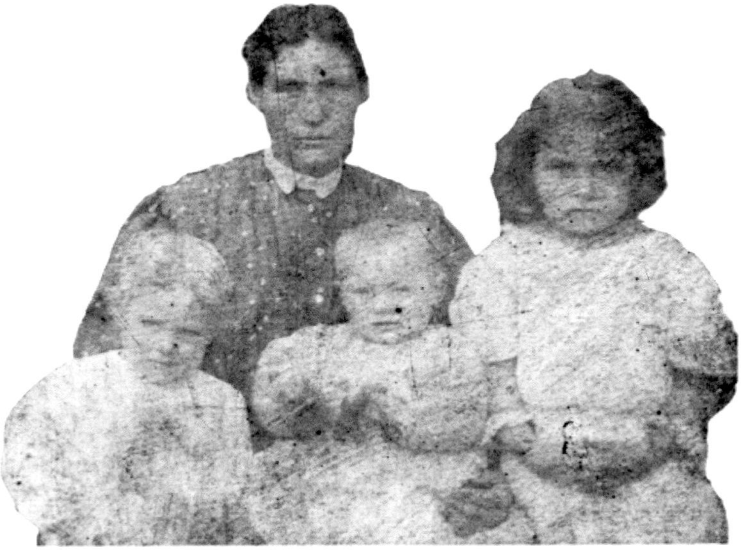

Figure 10-6. *With the background removed, the damage to the people is obvious.*

This is the point at which you can see how great the damage is. I've already rebuilt the eyes, but as you can see all the scratches, spots, and worn areas are obvious. Removing these is largely done one blemish at a time.

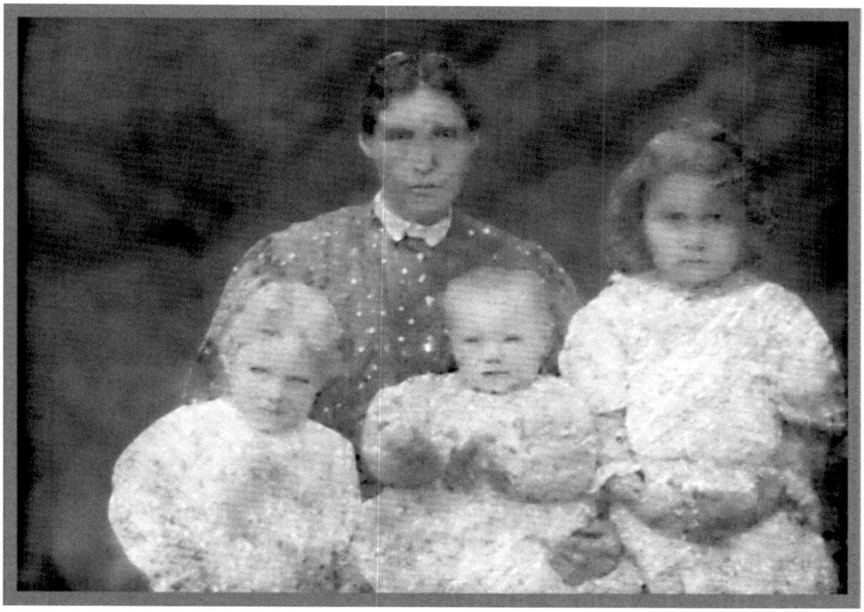

Figure 10-7. *Final version. NOT perfect, but a whole lot better.*

More handwork followed, refining eyes and mouths, cleaning up the dresses, etc.

Eventually, I blended the background and the figures, and gave the whole thing another round of nit-picking.

This work is not perfect. It would take another 3 hours or so to get there. But the client was totally delighted with it, and to spend 8 hours on extra coats of "hand rubbed lacquer" when the customer will only pay for enamel makes no sense.

The final image was printed on an ultra-fine grained fine art paper, and that's it.

The point of all this is that good restoration work is not something you run through automated stages. At every point you have a selection of tools to use, subroutines and plug-ins to call upon, and you have to take breaks or you go buggy from staring at the individual pixels too long.

Some of the tools used in restoring this picture were:

- Crop
- Levels
- BW Conversion
- Curves
- Blending Modes (multiply and pin light for sure)
- Noise Ninja
- Photoshop noise controls, especially median and dust-and-scratches)
- Spot healing
- Healing brush
- Patch
- Clone
- Pattern-make filter
- Dodge
- Burn
- Blur
- Sharpen

There is no one right way, no one right tool for all jobs. You just have to know what you're doing, and then use a fair amount of "undo" to get it all right.

WHAT TO DO WITH WATER DAMAGED PHOTOS

As a photo restorer, I see lots of water damage from floods, fire hoses, Windex, and years in a leaky attic. One of the most heart-rending losses is when photos get waterlogged, stuck to each other, or stuck to glass. Professional photo restoration experts can fix a lot of this damage, but there are number of things flood victims can do to help themselves.

The first thing to do is put your wet photos face-up on towels. Separate them; don't let them overlap or be in piles. The idea is to get air circulating around the pictures, so running a fan in the room will help as well. Keeping air moving reduces the amount of mold that forms on the picture. (Mold shows up as small spots, black, white, red, or green).

If you have photos that are stuck together, but only for a day or two, you can try this:

Get a jug of distilled water and a shallow pan (Tupperware is fine). Put the stack of pictures in the pan and cover with the water. Wait one hour and try to separate them. Let them soak another hour and try again. Finally, add ONE DROP of liquid dish soap to the pan, and go one more hour and try again.

What's happening is that there's a race between separating the pictures and lifting the actual images off the underlying paper, which is non-recoverable. So keep an eye on things.

The same technique works for pictures stuck to glass, but the soaking time may be longer. What I do when dealing with images stuck to the glass is clean the glass, scan the image through the glass, and THEN try separating. That way you have a back-up to work from.

If your picture has acquired a really nasty curve from drying unevenly, or being stored in a roll, here's a way to flatten it, at least temporarily. I actually did this with an 1898 8" x 48" panorama of a cavalry regiment, rolled to the size of a diploma for over 80 years. It worked, at least long enough to have a working copy made.

Take your picture, several towels, and two pieces of shelving bigger than the picture to the nearest Y or gym with a steam room. In the steam room put one towel on the bench, put the picture curl-side-up on the towel. Put the second towel on top, and crank up the steam. 15 to 20 minutes is probably enough to start the paper relaxing. At this point add a shelf to the stack, and add some additional weight if you want (sitting on the shelf might work, but ouch! those splinters!). Go into the steam room periodically (every five minutes is good), and when you judge your paper is as relaxed as it's going to get, turn the stack over and add the second shelf. Use large rubber bands or string to hold the whole assembly, and go on your way rejoicing -- as far as the nearest frame shop where you can have your image safely dry-mounted.

A LITTLE GHOST STORY

As some readers may know, I make much of my meager money from photo restoration. You can learn about my work in this area at hatchphotofix.com. My slogan for this work is "bring your memories back to life."

Earlier this week a client brought in an 8-by-40 school panorama shot in 1924. Huge panoramas were common in the early years of the last century. Schools, military regiments, even herds of cows were captured with special cameras that exposed enormous negatives up to 18 inches wide or even larger. The lenses in these specialized cameras were tack sharp, and the result (provided nobody moved causing motion blur) made faces amazingly clear.

My client's print was a school photo, with perhaps a couple of hundred kids in front of their school. The picture had a number of large cracks, some of them running through faces.

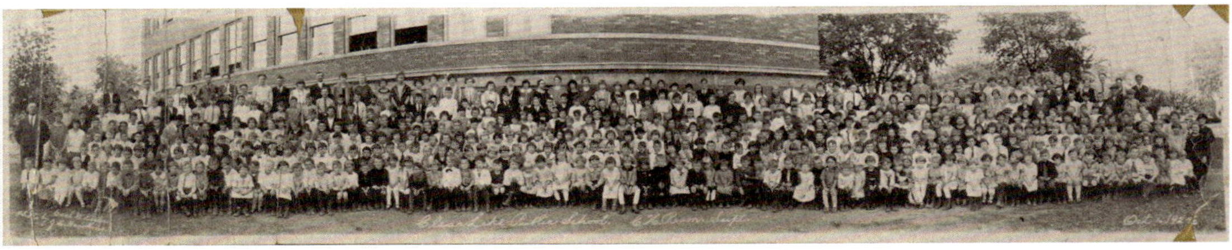

Figure 10-8. *40-inch wide panorama, ca 1925*

I winced when I saw this photo, because I've learned that restoring these large group images takes far more time than, say, a wedding portrait. That's because you really have to fix the faces, and the more there are, the harder they are to do. That makes panos of this sort expensive to restore, and few clients like the size of the estimate.

In this case, the important faces (to my client) weren't much damaged, so we agreed that to control costs, I'd do a minimal job on facial reconstruction, and deal with all the other problems (fading, missing areas, etc.) to a good-as-new standard.

To restore one of these monsters, you have to blow it up close to actual pixel size, work a bit, and then zoom out so you don't lose track of where you are in the picture. So I set to work, following the major cracks and repairing first the clothes (pretty easy) and then the faces (far from easy when a crack runs through an eye or a mouth – Figure 10-9).

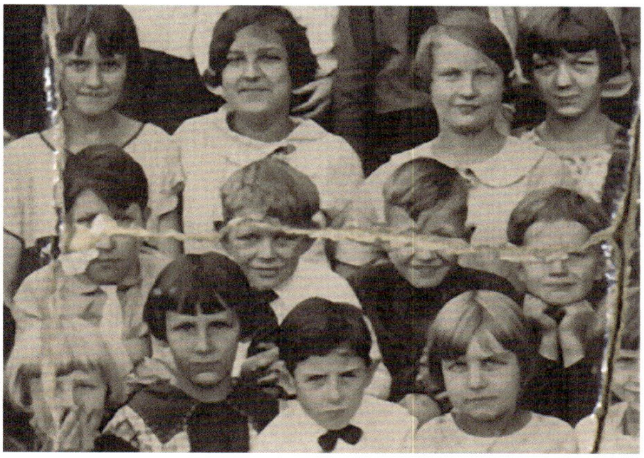

Figure 10-9. *You can see the damage plainly at full size.*

At first, I was only going to go for faces near the "important" ones (my client's relatives), and do a rough repair. And then it happened ... as I worked on these kids, their faces large enough and sharp enough to really see, I started to connect with them. Some of these kids were smiling, some were frightened, one is turned away. The older kids are at the rear, and they seem more relaxed, some joking around, some mugging for the camera.

As I worked, the children came to life and took hold of me. As I looked at each face from so long ago, I found myself wondering about the kid's life. Had she gone to college? Had he been killed in the roar of dive bombers on a Normandy beach? Why was this child grinning, and what made this little girl so fearful? One little boy in the front row looked so sad that I COULD NOT skip his face or just do a quick job. He wound up with an eye transplant and a new jaw line.

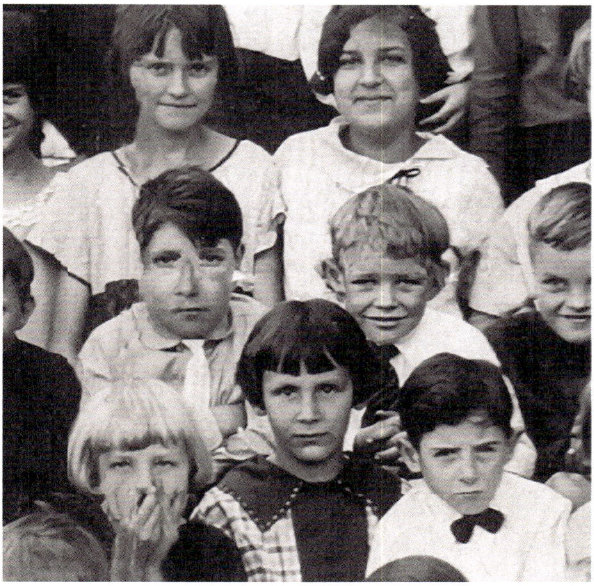

Figure 10-10. *Job done.*

The youngest of these kids would be 90 something now, so it's a good bet each of these little ones has passed on, perhaps unremembered by any one. It's an even better bet no other copy of this picture exists. And so it somehow became terribly important to give these damaged and forgotten children some kind of healing, some sort of life in this picture. They needed to look like themselves, as much as I could manage.

I worked to restore each wounded face, partly by thinking about them, partly by connecting with their expressions, partly by giving them this small gesture of appreciation for their youth, their feelings, and their lives now ended.

The work isn't perfect, but it's good. I hope it's good enough to show these little ghosts that someone cares.

Never before have I been so caught and held by the people in a restoration project. I don't know why it happened this time, except that the photographer somehow got it just exactly right, and the fine lens and film he used made it possible to really see each little face. Or maybe all those little ghosts reached out to me and asked for me to believe in them.

I know I'm not billing the client for all the time on this job. There's no need, because I feel that I'm the one who has been paid in experiencing whatever it is that bound me to these kids and brought them to life for me.

Sometimes it's not about the money.

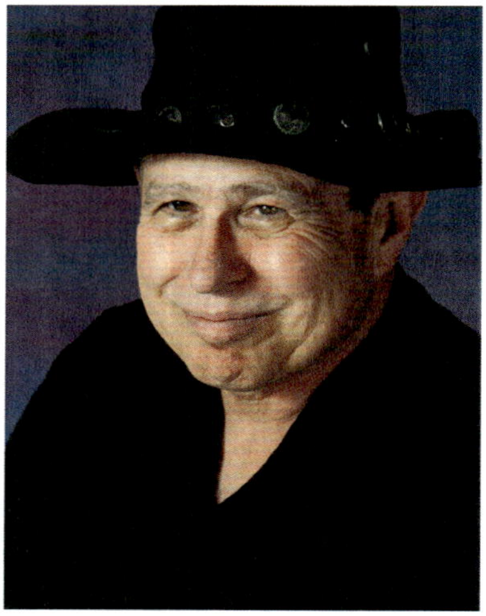

ERIC K. HATCH *has written over 80 articles, essays, speeches, features, and professional articles in the last 30 years. He has won two national awards: a Gold Quill from the International Association of Business Communicators, and Communicator of the Year from the Aviation/Space Writers Association.*

As a photographer, Eric has won numerous regional awards and competitions. His work has appeared in AAA magazine, Oxygen Magazine, Bicycling, Alaska Milepost (annual) and Wooden Boat, among others. He has served on the board of the Southern Ohio Professional Photographers Association. He focuses on travel and fine art photography and specializes in digital photo restoration.

Made in United States
Orlando, FL
04 November 2026

90202166R00140